Contents

Peanut Health Management

F. M. Shokes
North Florida Research and Education Center
University of Florida, Quincy

H. A. Melouk
U.S. Department of Agriculture
Agricultural Research Service
Stillwater, Oklahoma

CHAPTER ONE

Plant Health Management in Peanut Production

Growing totally unblemished or healthy plants can be accomplished only in a pathogen-free and stress-free environment. Therefore, growing plants under normal field conditions exposes them to a physically and biologically hostile environment that affects plant health adversely to varying degrees. Plant health refers to the absence of disease, injury, or defects in individual plants or a population of plants. It relates to the general condition of an individual plant or a crop (i.e., a population of plants) in a particular environment. To evaluate plant health we must consider all parts of a plant and every aspect of its growth. A sound, well-developed root system and a sturdy system of stems or branches to support the foliage are necessary for the good health of most plants. Foliage must be healthy to sustain adequate photosynthesis, and flowers must be in good condition for seed production. Requirements for the development of these will vary with different types of plants. Any given plant species is adapted for growth within certain environmental conditions. Therefore, the climate and the soil environment are important in any discussion of plant health.

Plant health is directly important to those who grow plants for food or fiber. Though they are often unaware of it, plant health is of extreme importance to all consumers of plant products. Today there is an increasing interest in food safety and preservation of the environment. Therefore, any plant health management scheme must be designed for production of safe, high-quality products. It should use systems that are economically sustainable and environmentally sound.

Scope and Objectives of This Book

This book covers all principles and aspects of growing a healthy peanut crop. It is intended especially for peanut growers, extension personnel, students with interest in this unique crop, agricultural and food industry personnel, and those involved in peanut research. Because of the varied interests of those using this book, it is structured to meet different needs. Much of the text is nontechnical. You may read and enjoy it without getting bogged down in an excess of technical jargon of specialized disciplines. Important information is highlighted in boxes to allow quick scanning of topics for essential facts and principles without the details. When more detail is required, it may be obtained by closer reading of the text.

The main objective of this book is to present basic concepts and strategies for growing and managing a healthy peanut crop. Discussion of these concepts and strategies can help in making informed management decisions while taking into account economic realities and protecting the environment. To meet this objective we visualize growing a peanut crop as a building process, like building a pyramid on a strong base. Building a strong base is essential in growing a healthy crop. Thus, discussion of concepts relating to building a strong base, such as site selection, soil fertility, selection of variety or cultivar, seed quality, and plant growth and development, are integral parts of this book. Several chapters are devoted to strategies for managing all types of peanut plant health problems, including practices for managing weeds, insects, mycotoxins, and physiological and environmental disorders and diseases. One chapter covers pesticide application techniques in detail. Finally, a treatise such as this must also include the bottom line—economics—and thus a chapter is included on the economic aspects of plant health as it relates to management choices. The characteristics of this unique plant have been considered in the development of management schemes for peanut production.

The Peanut Plant

Peanut belongs to the genus *Arachis*, which includes more than 20 described species of tropical and subtropical plants in the legume family. The cultivated peanut belongs to the

1

species *Arachis hypogaea* L., which does not usually occur in the wild.

Peanut is an upright or prostrate annual legume that is sparsely hairy. It generally grows 6–24 inches (15–60 centimeters) high and produces a well-developed taproot with many lateral roots (Fig. 1.1). Peanut leaves are alternately arranged on the stems. Each leaf has four leaflets, about 1–4 inches (2–10 cm) long, in opposing pairs. The yellow flowers (Plate 1) are located in the axils of leaves at nodes not occupied by branches. One to many flowers may grow from a node, with the greatest abundance at the lowest nodes.

Flowers begin to appear 4–6 weeks after planting; the greatest number appear 6–10 weeks after planting. The flowers are self-pollinated around sunrise and wither within 5–6 hours. Within 1 week of fertilization, a pointed carpophore, or gynophore, commonly known as a peg, develops and elongates (Plate 1). The fertilized ovary is located behind the tip of the peg. The peg responds to positive geotropism; that is, it grows downward to enter the soil. There it loses its geotropism and becomes oriented horizontally as the ovary enlarges and pod growth begins.

Mature pods may contain one to five seeds. The testae, or seed coats, vary in color depending on the cultivar but are typically tan, pink, or red. Seed weights vary from 0.2 to 2.0 grams depending on the market type (U.S. market types include Spanish, runner, and Virginia). Spanish peanuts have small seeds, Virginias have large seeds, and runner peanuts have seeds of intermediate size. Each seed consists of two large cotyledons (embryonic leaves), an epicotyl, and a primary root.

Box 1.1 contains definitions of terms used in this chapter.

The growth stages of the peanut plant are based on visual observation of the vegetative and reproductive events. The growth stages of peanut are presented in Table 1.1. The repro-

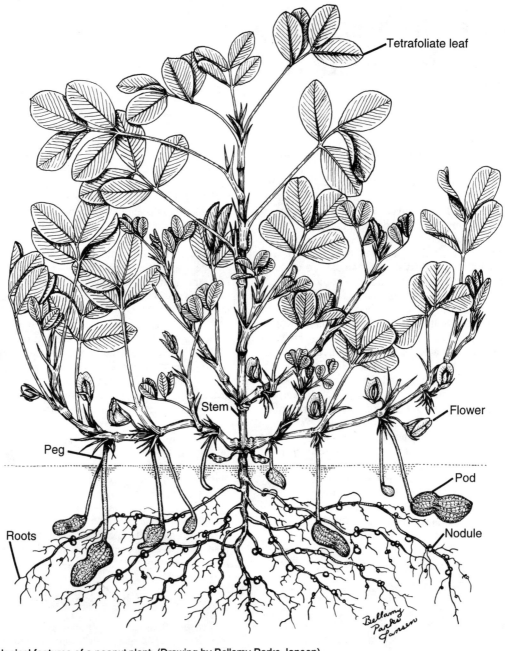

Fig. 1.1. Morphological features of a peanut plant. (Drawing by Bellamy Parks Jansen)

ductive stages of a runner peanut are shown in Plates 1–9.

There are two major subspecies of *A. hypogaea: hypogaea* and *fastigiata* (Table 1.2). The subspecies *hypogaea* has alternate branching and either a spreading or a bunching growth habit. Seed of this subspecies undergoes a dormancy period after maturity. Within this subspecies are two botanical varieties. The Virginia and runner market types are in the botanical variety *hypogaea,* and the Peruvian humpback or Chinese dragon type is in the botanical variety *hirsuta.* Plants in the subspecies *fastigiata* have sequential branching, an erect growth habit, relatively short maturity time, and little or no

seed dormancy. This subspecies is also divided into two botanical varieties. The Valencia market type corresponds to the botanical variety *fastigiata,* and the Spanish market type corresponds to the botanical variety *vulgaris.*

Peanut is believed to have been first domesticated in Paraguay. Remnants of peanut have been found at archaeological sites dating from 1500 to 1200 B.C. on the northern coast of Peru. Currently, the crop is grown on all continents in more than 50 countries. Commercial crops predominantly are found in tropical or temperate regions of the world. China, India, and the United States are the leading producers of peanut (Table 1.3). Since peanut has a high oil content (50%), a large percentage of the world production is used as oil for cooking or salads. About 70% of the production in the United States is used in domestic food products, including peanut butter, salted products, confections, and roasted nuts. The remainder is used for seed or oil or is exported. In the United States, peanut is produced on about 1.7 million acres, concentrated mainly in three major geographic areas: the Southeast, which includes Alabama, Florida, Georgia, and South Carolina; the Southwest, which includes New Mexico, Oklahoma, and Texas; and Virginia-Carolina, which includes North Carolina and Virginia (Table 1.4; see also the map in the color section).

Major Plant Health Problems in Peanut Production

When considering plant health, we must take into account all parts of the plant and the problems encountered during its growth and development. The peanut plant begins as a seed, and as such it is affected by environmental conditions, soil-borne pathogens, and preharvest and postharvest insects. Soil fertility is critical, especially calcium (Ca) and potassium (K) levels, and can be a major factor in producing high-quality seed.

Development of a healthy root is also important. Because peanut is a legume, its root system must be exposed to the proper strain of *Bradyrhizobium* bacteria (nitrogen-fixing bacteria). Effective nodulation by the bacteria must occur for

---

Box 1.1

Definitions

bunching growth habit a clustering or tufting growth and appearance

cotyledon the primary leaf of the embryo of a seed plant

crown the point where the stem and root join, at or just below the surface of the ground

dormancy a period of inactive or still growth

endosperm the nourishment surrounding the embryo in the seed

epicotyl the part of the stem of a seedling just above the cotyledon

gynophore (peg) a stalk bearing the fertilized female organ of a flower

node the part of a stem from which a leaf or a shoot branch starts to grow

ovary the enlarged hollow part of the pistil in a flower

pericarp the wall of a ripened ovary

pod the seed case of peanuts

spreading growth habit a scattering or diffusing, extending growth and appearance

testa (pl. testae) the skin or coat of a seed

tetrafoliate having four leaflets

Table 1.1. Growth stages of peanut[a]

	Abbreviated title of stage	Description
Vegetative stages[b]		
VE	Emergence	Cotyledons near the soil surface, with the seedling showing; some part of the plant visible
VO		Cotyledons flat and open at or below the soil surface
V-1 to V-(N)	First tetrafoliolate to Nth tetrafoliolate	One to N developed nodes on the main axis (a node is counted when its tetrafoliolate is unfolded and its leaflets are flat)
Reproductive stages[b]		
R1	Beginning bloom	One open flower at any node on the plant
R2	Beginning peg	One elongated peg (gynophore)
R3	Beginning pod	One peg in the soil with turned, swollen ovary at least twice the width of the peg
R4	Full pod	One fully expanded pod, to dimensions characteristic of the cultivar
R5	Beginning seed	One fully expanded pod in which seed cotyledon growth is visible when the fruit is cut in cross section with a razor blade (past the liquid endosperm phase)
R6	Full seed	One pod with cavity apparently filled by the seeds when fresh
R7	Beginning maturity	One pod showing visible natural coloration or blotching of the inner pericarp or testa
R8	Harvest maturity	Two-thirds to three-fourths of all developed pods showing coloration of the pericarp or testa (the fraction is cultivar-dependent, being lower for Virginia types)
R9	Overmature pod	One undamaged pod showing orange-tan coloration of the testa or natural peg deterioration

[a] Source: Boote, 1982.

[b] For populations of plants, vegetative stages can be averaged if desired. Reproductive stages should not be averaged. A population of plants remains at a given reproductive stage until the date when 50% of the plants in a sample have developed the distinctive trait of the next stage. An individual plant enters a new reproductive stage at the time of the first occurrence of the distinctive trait of that stage, without regard to position on the plant.

fixation of atmospheric nitrogen to usable forms. Nitrogen (N) deficiency can occur if the proper bacterial strain is not present. This is just one of many problems related to soil fertility. During the early stages of growth, roots are susceptible to infection by several soilborne pathogens. As the tap-root extends downward and secondary roots spread outward, they will encounter various soil fungi, nematodes, and insects. Any organism or environmental condition that is detrimental to the roots will impair their ability to absorb moisture and nutrients, thus affecting the entire plant. The level of soil

Table 1.2. Subspecies of *Arachis hypogaea*

Subspecies	Site of flowering and pod production	Growth habit	Botanical variety and market type	Seed dormancy	Maturation time
hypogaea	Lateral branches	Spreading Bunching	*hypogaea*, runner *hypogaea*, Virginia *hirsuta*, Peruvian humpback or Chinese dragon	Present Present	Long (145–165 days)
fastigiata	Main stem	Erect	*fastigiata*, Valencia *vulgaris*, Spanish	Low or absent	Short (125–145 days)

Table 1.3. Peanut area, yield, and production[a]

	Area (million hectares)		Yield (metric tons per hectare)		Production (million metric tons)	
	1992	Nov. 1993	1992	Nov. 1993	1992	Nov. 1993
World	19.34	19.86	1.19	1.15	23.08	22.74
United States	0.68	0.67	2.87	2.21	1.94	1.48
India	8.39	8.55	1.03	0.87	8.60	7.40
China	2.98	3.25	2.00	2.22	5.95	7.20
Indonesia	0.66	0.67	1.48	1.48	0.97	0.99
Senegal	0.88	0.88	0.82	0.82	0.73	0.73
Myanmar (Burma)	0.48	0.54	0.89	0.85	0.43	0.46
Argentina	0.12	0.12	2.39	2.50	0.28	0.30
Sudan	0.55	0.55	0.71	0.71	0.39	0.39
Zaire	0.53	0.53	0.72	0.72	0.38	0.38
Nigeria	0.50	0.50	0.50	0.50	0.25	0.25
Vietnam	0.30	0.30	0.98	0.98	0.30	0.30
South Africa	0.16	0.15	1.05	1.07	0.17	0.16
Brazil	0.09	0.09	1.69	1.67	0.15	0.15
Thailand	0.12	0.13	1.32	1.32	0.16	0.17
Burkina Faso	0.23	0.23	0.69	0.69	0.16	0.16
Central African Republic	0.13	0.13	1.12	1.12	0.15	0.15
Cameroon	0.32	0.32	0.44	0.44	0.14	0.14
Côte d'Ivoire	0.15	0.15	0.98	0.98	0.15	0.15
Gambia	0.10	0.10	1.26	1.26	0.12	0.12
Uganda	0.14	0.14	0.79	0.79	0.11	0.11
Others	1.86	1.88	0.85	0.85	1.57	1.59

[a]Source: Production Estimates and Crop Assessment Division, Foreign Agriculture Service, U.S. Department of Agriculture, November 1993.

Table 1.4. Acreage, yield, and production of peanut by region in the United States[a,b]

	1990	1991	1992	1993
Harvested area (1,000 acres)				
Southeast	1,133.5	1,304.0	1,002.0	1,010.5
Southwest	415.0	453.7	424.2	382.8
Virginia-Carolina	261.0	258.0	246.0	244.0
United States	1,809.5	2,015.7	1,672.2	1,637.3
Yield (pounds per acre)				
Southeast	1,762	2,439	2,641	2,031
Southwest	1,976	2,154	2,297	2,241
Virginia-Carolina	3,010	2,980	2,696	1,703
United States	1,991	2,444	2,562	2,032
Production (1,000 pounds)				
Southeast	1,997,285	3,180,295	2,646,545	2,053,015
Southwest	819,970	977,375	974,566	857,840
Virginia-Carolina	785,515	768,900	663,195	415,645
United States	3,602,770	4,926,570	4,284,306	3,326,500

[a]Source: U.S. Department of Agriculture, Economic Research Service, January 1994.
[b]Southeast includes Georgia, Florida, Alabama, and South Carolina. Southwest includes New Mexico, Oklahoma, and Texas. Virginia-Carolina includes Virginia and North Carolina.

moisture is important, and drought conditions or excessive water may affect the ability of the plant to grow, reproduce, and produce acceptable yield.

Aboveground, the crown area of the plant stem is subject to attack by insects and soil microorganisms, such as fungi. Because plant tissue is tender during the seedling stage, organisms such as the fungus *Aspergillus niger* may attack the plant and cause a crown rot. As the plant approaches maturity and foliage produces a nearly complete ground cover, the plant is subject to stem rot and limb rot diseases. Injuries to foliage caused by insects or disease reduce the ability to produce seed. Leaf spot diseases, for example, will defoliate peanut plants and severely decrease their ability to produce acceptable pod yields. Plants can be attacked by numerous foliage-feeding insects. Environmental stress, such as extreme temperature, may be a problem at certain times of the growing season. Weeds must be controlled or they will compete with peanuts for light, moisture, and nutrients and may interfere with foliar disease management. Fungicides applied to peanut foliage may be intercepted by weeds, thus increasing the possibility of disease problems.

Because peanut pods are produced on pegs, the functional condition of this plant part is important for peanut health. As pegs grow downward into the soil, they are exposed to the same soilborne organisms that can affect the root and crown. Once the pegs are in the soil, moisture and nutrients are critical for proper pod development. When pods are exposed to soilborne pathogens and pod-feeding insects, they may be injured, rotted, or simply blemished. Under moisture stress, pods may be predisposed to infection by mycotoxin-producing fungi, such as *Aspergillus flavus*. This type of contamination can be influenced by harvesting, handling, and storage practices.

The Concept Behind Strategies for Peanut Health Management

Growing a healthy peanut crop may require somewhat different practices in different growing regions. However, the basic idea used to develop a crop management strategy is that of providing an environment that allows maximum yield with reduced risks of loss, proper use of pesticides and other petrochemicals, and minimal environmental contamination. A successful plant health management strategy must include management of all the following:

1. Physiological and environmental disorders
2. Weeds
3. Preharvest and postharvest insects
4. Viral diseases
5. Foliar pathogens
6. Soilborne fungal pathogens and nematodes
7. Mycotoxin-producing fungi

Pesticide application techniques must also be considered, since wise use of pesticides plays a role in five of these seven peanut health problems.

Integration of Management Strategies

To be practical at the farm level, plant health management strategies must be integrated into a package approach. For this reason, we have assembled guidelines that include most of the commonly used management strategies for each peanut-growing region or state. One example of how management practices overlap is the practice of deep plowing of soil to turn under debris and inoculum of the stem rot pathogen, *Sclerotium rolfsii;* this also helps to reduce the available inoculum of the leaf spot pathogens that can survive on leaf and stem debris. For any health management practice to be practical, it must complement other crop management strategies. At appropriate places we have included sections dealing with integrating management practices into the entire growing package. A suggested timetable for peanut production in the United States is presented in Box 1.2.

Box 1.2

Suggested Timetable for Peanut Production in the United States

Activity	Chapter	Region[a] Southwest	Region[a] Virginia-Carolina	Region[a] Southeast
Land preparation and soil fertility	2	April–June	February–May	February–May
Planting	3 and 4	April–July[b]	April–May	April–May
Irrigation	3	June–November	June–September	June–September
Management decisions to control				
Foliar diseases	10	May–October	June–September	May–September
Soilborne diseases	11	April–September	April–August	April–August
Viral diseases	9	April–July	April–May	April–May
Arthropods	8	April–October	April–August	April–September
Weeds	7	April–July	April–May	April–July
Nematodes	12	April–July	April–May	April–May
Aflatoxins	13	June–December	May–October	May–October
Harvesting and curing	5	July–November	September–November	August–November

[a]Southwest includes New Mexico, Oklahoma, and Texas. Virginia-Carolina includes Virginia and North Carolina. Southeast includes Georgia, Florida, Alabama, and South Carolina.
[b]Planting in South Texas begins in February.

Selected References

Boote, K. J. 1982. Growth stages of peanut (*Arachis hypogaea* L.). Peanut Sci. 9:35-39.

Pattee, H. E., and Young, C. T., eds. 1982. Peanut Science and Technology. American Peanut Research and Education Society, Yoakum, TX.

Porter, D. M., Smith, D. H., and Rodríguez-Kábana, R., eds. 1984. Compendium of Peanut Diseases. American Phytopathological Society, St. Paul, MN.

Shokes, F. M., and Smith, D. H. 1990. Integrated systems for management of peanut diseases. Pages 229-238 in: Pest Management in Agriculture, 2nd ed. Vol. 3. D. Pimentel, ed. CRC Press, Boca Raton, FL.

F. R. Cox
Department of Soil Science
North Carolina State University, Raleigh

J. R. Sholar
Department of Agronomy
Oklahoma State University, Stillwater

CHAPTER TWO

Site Selection, Land Preparation, and Management of Soil Fertility

Important factors in growing a healthy, high-quality, and profitable peanut crop are the proper choices of cultural practices, which include site selection, crop rotation, land and seedbed preparation, and management of soil fertility. These choices are generally based on experience, available literature, and consultation with individuals knowledgeable about peanut culture.

Site Selection

Selecting fields with a suitable type of soil is essential for profitable peanut production. Peanut grows best in deep, well-drained soils with a sandy or very friable (loose) surface layer. Moist soil rubbed between the index finger and thumb should not ribbon out but should fall apart easily. In soils with good depth (24–30 inches) and drainage, peanut will develop an extensive root system, making the crop less susceptible to drought damage. Good drainage also means good aeration (exposure to air and gases), which is essential for a legume such as peanut to adequately fix nitrogen (N) and for roots to grow well.

Organic matter (the nonelemental component of the soil) is another factor associated with good drainage. In addition, soils with low (less than 2%) organic matter are less likely to harbor certain insects and disease-causing microorganisms than those with high organic matter levels. Also, high organic matter levels may contribute to pod discoloration. Organic components in the soil may actually stain the pods, and any such discoloration would be detrimental to the crop, especially if it is to be sold in the shell.

Peanut ordinarily is grown on rather sandy soils for several reasons. 1) Sandy, well-drained soil generally does not lead to pod discoloration caused by either organic matter or iron oxides. 2) The peg is able to penetrate sandy soil easily. Unless it is extremely friable, a soil with a greater clay content (fine textured) tends to crust more than sandy soils when dry, making penetration more difficult for the pegs. 3) Pod size is reduced if the soil is fine textured. 4) Harvesting is easier be-

cause sandy soils facilitate the digging process. Excessive clay tends to adhere to pods, which may cause the peg to break and the pod to be lost and remain in the soil. When dry, soils with high clay content can form clods. If these clods are similar in size to pods, they may not be cleaned out of the combine at harvest. For the same reason, rocky soils should not be used for peanut production.

The use of very sandy (more than 80%) soils for peanut production, however, has its disadvantages. These soils hold less water and thus are more prone to drought. Peanut needs a good supply of water (22–25 inches), well distributed throughout the growing season, to produce economical yields (2,500 pounds or more per acre). Sandy surface soils are particularly prone to dry out quickly. Such drying during fruiting restricts

> **Selecting fields with a suitable type of soil is essential for profitable peanut production.**

the flow of nutrients, particularly calcium (Ca), to the pods. Sandy soils also have a low cation exchange capacity, so their fertility must be managed very closely. Lime must be applied more frequently on acid soils. Tests of sandy soils may indicate a low supply of potassium (K), but there is usually a sufficient amount in the subsoil, especially if there is an accumulation of clay and prior crops have been fertilized. Levels of some micronutrients, especially boron (B), will also be lower in the more sandy soils. Fertility of sandy soils must be handled more carefully than that of soils with more clay and higher exchange capacities.

Clay soils may be used for peanut production only if they have a good granular structure (friable soils). Pegs enter friable soils readily, and pods can be harvested cleanly. Such soils must also be well drained to be acceptable for growing peanut.

Crop Rotation

Rotating peanut with other crops, such as grasses, sorghum, corn, and cotton, can be beneficial in several ways. The term "rotational effect" has been used to describe the increase in crop yield that occurs when crops are rotated compared with yields obtained by continuous cropping under similar conditions. Several factors contribute to this effect when peanut is grown in an appropriate rotation: 1) more effective use of residual soil fertilizer; 2) improved efficiency in controlling certain weeds; and 3) reduction in soilborne disease and nematode problems.

The taproot of the peanut plant can reach a depth of 6 feet, which makes it very effective in utilizing residual fertility. The most common practice for meeting the nutrient requirements for peanut has been to appropriately fertilize the preceding rotational crop.

Crop rotation is primarily used as a cultural method for reducing the effects of pests on peanut health. Because peanut is a high-value cash crop, economic considerations are likely to influence the decision to rotate peanut with other crops. A 3-year rotation, where peanut is planted every third year, can increase peanut yield by about 30%. Additional comments on the value of crop rotation in peanut production are included in several chapters of this book.

Land Preparation

Conventional land preparation for peanut production usually involves primary and secondary tillage operations to obtain a friable seedbed. Primary and secondary tillage are performed to control diseases and weeds and to prepare a seedbed for planting. Soil preparation begins with deep turning of the soil with a moldboard plow followed by secondary tillage (disking) to provide a seedbed that is relatively smooth, weed free, and almost free of crop residues. There are new designs in plows, such as the switch plow, which bury crop residues better than the conventional moldboard. Studies have shown that no-tillage or less intensive tillage systems have generally produced lower yields than systems that use a moldboard plow. The practice of disking after moldboard plowing has been beneficial in reducing compaction, soil erosion, and land-preparation costs.

Fall plowing is rarely practiced in any of the peanut-growing areas unless the cultural plan includes seeding a winter cover crop. Plowing very early in the season results in higher yields than plowing just before planting. Early plowing can reduce the incidence of several soilborne diseases, such as white mold or southern stem rot (Chapter 11), and possibly even the inoculum of some foliar pathogens (Chapter 10).

The use of the moldboard plow and other implements on sandy soils can lead to the development of a plowpan, a hard, brittle horizon below the plowlayer. This layer will almost completely restrict the growth of roots of some crops and does restrict the growth of peanut roots. Subsoiling (deep ripping of the soil) can break up the plowpan, and sometimes peanut fields are subsoiled beneath the rows. There have been occasional positive responses to subsoiling, mainly when the plowpan is severe and the climate is extremely dry. Subsoiling should help roots develop more deeply into the profile and utilize more subsoil moisture.

Peanut also grows well on raised beds, which may contain a single row or a pair of rows. Planting on a bed improves drainage, thus warming the soil and improving aeration. Both warm temperatures and a good oxygen supply are important

for peanut growth (Chapter 3). Peanut should not be planted until the soil temperature is at least 65°F (about 18°C) and is expected to continue at that level for a few days. A good oxygen supply is essential not only for root growth, but also for N fixation.

Planting peanut on a bed has other advantages. If the soil has to be cultivated (worked for removal of weeds), there is less possibility of burying part of the plant. The crop can be dug more easily; the soil is not as likely to be too wet; and it should not be as difficult to keep the digger blades at the correct depth.

The land-preparation method may have to be altered because of fertilization or liming practices. If K must be applied directly before peanut is grown, it should be incorporated in such a manner that it does not compete with Ca in the pegging zone. This means it should be applied before moldboard plowing and turned down so that most of it is 4–8 inches (10–20 centimeters) deep. This has also been the practice for liming, but now it is felt that applying lime after plowing, by disking it in, will provide extra Ca in the pegging zone for pod development.

Concerns about soil loss to wind and water erosion and about large energy requirements for conventional tillage systems have increased interest in conservation tillage for peanut production. Conservation tillage systems involve producing small grains during winter months and then planting peanut in the small grains residue. Small grains residue left on the soil surface during critical erosion periods can significantly reduce erosion.

Conservation tillage systems fall into four general categories: 1) no-till, seeding directly into previously undisturbed soil; 2) minimum tillage, performing the least soil disturbance necessary for crop production or for meeting tillage requirements under existing soil conditions; 3) reduced tillage, planting in a system that consists of fewer or less energy-intensive operations compared with conventional tillage; or 4) strip tillage, planting in a system in which 30% or less of the soil surface (bands in the row) is tilled.

Research and field experience have demonstrated that conservation tillage practices can be used in peanut production. However, surface or subsurface tillage promotes peanut seed germination and growth. Pod yields from minimum, reduced, or strip tillage systems are generally lower than yields obtained with conventional tillage systems. Soil compaction, uncontrolled weeds, and poor stands in no-till systems may result in reduced yields. Yields from no-till systems are consistently lower than yields from conventional or minimum, reduced, or strip tillage systems.

Soil Fertility

Nitrogen Fixation

Peanut is generally not responsive to the application of N, phosphorous (P), or K fertilizers if the soil has been well managed for previous crops. Because the plant is a legume, it fixes N if a few conditions are met. One condition is that the proper strain of *Bradyrhizobium* bacteria (cowpea group) must be present to infect the roots and cause nodule formation. The cowpea group of *Bradyrhizobium* is effective in nodulating a broad spectrum of plant species. In many places, these bacteria are common enough in the soil that no response is obtained to inoculation of the plant or infestation of the soil. It is a general practice, however, to apply inoculant if peanut has not been grown in a field for several years. Several formulations of the bacterial inoculant can be purchased from peanut seed dealers, farmer cooperatives, or agribusiness stores.

The most effective method of inoculating seed is to apply a granular inoculant in the furrow. The granular inoculant is applied at a much higher rate than seed inoculants used in the past. The granular method, therefore, not only provides greater numbers of bacteria, but also positions them to be readily available to developing roots.

The range of soil conditions under which fixation of N should be sufficient is fairly broad. However, a prolonged dry season can limit N fixation. In areas with extreme dry conditions, the viability of the rhizobia is affected so much that infection is reduced. Another factor is soil acidity. At pH levels below 5, Ca may become limiting, and the availability of molybdenum (Mo), which is needed by the enzymes in the N fixation process, will be decreased.

If the correct type of rhizobia is lacking or if conditions are poor (because of drought, for example) for N fixation by the bacteria, there will be a deficiency of N available to the plant (Plate 10). Although there will be a lack of growth from insufficient amino acids and proteins, the first striking symptom of N deficiency is general chlorosis. Nitrogen is an integral part of chlorophyll, and with less chlorophyll, there will be less green color. As a result, the crop will be lighter in color or even turn slightly yellow. The deficiency symptom will occur quite uniformly across the entire plant in moderate to acute stages. Early N deficiency may appear in the lower leaves first.

When N fixation is restricted, symptoms may not be observed while the plant is in the vegetative stage (Chapter 1). When the plant is in the middle to latter part of the reproductive stage (Chapter 1), however, symptoms usually will appear. This is because there is a tremendous amount of translocation of N from the leaves and stems to the developing fruit during this stage. Conversely, if fruiting is restricted, for example, by a lack of Ca, then the plants may be abnormally green late in the season.

Aluminum

Peanut is susceptible to aluminum (Al) toxicity caused by low soil pH. Root growth is reduced and yields are affected, so the general recommendation is to maintain the pH above 5.8. If extractable Al is not excessive, however, peanut may be grown at lower pH levels, even as low as pH 5.0. Keeping the pH level high retains cations in the topsoil, especially Ca, and eventually improves conditions in the subsoil.

Another reason for keeping the soil pH well above the minimum requirement is that pH is affected by other fertilization practices. Any addition of salt, such as from a fertilizer, will cause a temporary decrease in the pH. With peanut, it is common to apply gypsum or landplaster (calcium sulfate), which causes a temporary decrease in soil pH, usually of about 0.5 units. Depending on rainfall thereafter, this drop may last a month or two, and then the pH returns to normal. Soil tests are very useful and are recommended for determining the lime requirement to keep the pH in the correct range.

Plant symptoms of excessive acidity or Al toxicity will be most pronounced in the roots. Growth is retarded, and roots will be stubby and sparse. Growth of the tops will naturally be inhibited somewhat. Excessive Al will also decrease the uptake of Ca. Symptoms of Ca deficiency will be discussed later.

Phosphorus

Soil tests are also recommended for determining the P level in the soil in relation to requirements for peanut production. Phosphorus is adsorbed tightly by the soil, so there is essentially no leaching. There is some very tight sorption (adherence), sometimes called fixation, but this tends to act as a reserve. Continued applications of P fertilizer have generally built up not only the P levels extractable by soil tests, but also the level in the reserve. As a result, there are very few P deficiencies of peanut in the United States.

The point at which peanut becomes deficient in P on the basis of the soil test is called the critical level. This level depends upon the extractant and the soil test used. The critical levels of P in three methods, Mehlich-1, Mehlich-3, and Bray-1, are about 10, 20, and 20 milligrams per liter, respectively. Multiplying these numbers by two will approximate the level in kilograms per hectare (kg/ha), which is also roughly pounds per acre (lb/acre). The critical levels for other crops grown in rotation with peanut, such as corn, are almost always greater than these. As a result, if the other crops in the rotation are fertilized correctly, there should be no need to fertilize peanut with P directly. Since peanut is a high-value crop, however, considerable fertilizer is applied under the guise of "insurance," even though no response is expected.

Phosphorus deficiencies are seldom severe enough to produce distinct symptoms in peanut plants. Plants will be stunted, and leaf size will be reduced. With the reduced growth, the color may actually become darker. In extreme cases, a reddish color may develop in both leaves and stems.

Potassium

Potassium fertilization may also be based on soil tests, but soil tests are less reliable for determining the need for K than for P. This is because the cation exchange capacities of soils on which peanut is grown are ordinarily quite low, so there is minimal retention of K in the topsoil. Potassium will leach to lower horizons, and it will be somewhat fixed in certain clay minerals usually present in subsoils. The leaching of K is promoted by not keeping the soil pH in the optimum range and by applying gypsum or landplaster.

Peanut has an extensive root system and, as such, can utilize K in lower horizons of the soil. It is a common and recommended practice, therefore, to simply be sure the previous crop is fertilized adequately and then let the peanut crop use the residual K. If K is applied directly prior to planting, it is suggested that it be plowed down. This is to place it below the pegging zone, since high K in the pegging zone interferes with Ca uptake by the pegs and pods.

Symptoms of K deficiency in peanuts, like those of P, are seldom severe. Mild deficiencies result in some stunting of the crop and reduced yield. More severe deficiencies result in a chlorosis that begins at the edges of the leaves and gradually works its way inward. The leaves may become entirely brown and drop off, especially the older leaves (Plate 11).

Calcium

The extra need for Ca is unique to the peanut among crops and is created by the fruiting pattern and soil conditions. The Ca taken up by roots is translocated up the stems to the leaves, but very little of it is translocated back down the pegs to the developing fruit. The Ca required by the fruit, therefore, must be taken up directly by the fruit. The ability of the pod to take up sufficient Ca from the soil depends upon the type of peanut being grown. Virginia types are less able to take up adequate Ca than runner and Spanish types. This may simply be a matter of pod size, since there is less surface area on larger pods per unit weight of nut.

Soil conditions also affect the availability of Ca. As with K, the availability of Ca may be limited because peanut is often grown on soils with very low cation exchange capacities. Soil tests are used to determine the need for fertilization with Ca for the runner peanut. The soil test critical level is 300–500 pounds of Ca per acre (about 340–570 kg/ha). Preliminary results with Virginia type peanuts indicate that the critical level should be at least 1,500 lb/acre (about 1,700 kg/ha); but since many soils have such a low exchange capacity, this level may never be attained, and a recommendation for extra Ca is made across the board for all Virginia type peanuts.

The usual source of Ca applied to the Virginia peanut is calcium sulfate, $(CaSO_4) \cdot 2H_2O$. This material is mined directly in some locations and there are several by-product forms, so the concentration of actual $CaSO_4$ may vary from 50 to 90%. Suggested rates of $CaSO_4$ are 250–600 lb/acre applied in a broad band about 18 inches wide over the row at early flowering. Band application is suggested because the material is needed only where the fruit are developing, and the timing is suggested because the extra Ca is needed only during fruiting. Earlier application is not recommended because this form of Ca is subject to leaching. Some materials are broadcast over the soil, in which case the rate of $CaSO_4$ is adjusted according to the soil surface area covered.

Calcium sulfate is also used on runner peanut. If the soil pH needs adjusting, however, lime may be worked a few inches into the surface, or fruiting zone, to provide additional Ca for runner peanut. Calcitic lime is better than dolomitic for this purpose, but the magnesium (Mg) needs of other crops in the rotation should also be considered.

The roots and tops of peanuts seldom, if ever, show Ca deficiency symptoms (Plate 12). Symptoms are found almost entirely in the developing fruit, either in the complete lack of fruit formation (in which the tip of the peg dies) or in various degrees of incomplete fruiting. With a trace of Ca, the shell may develop but kernels fail to form, resulting in an empty pod, or "pop." With slightly more Ca, perhaps only one of the usual two kernels will form, and it may not be normal. If a kernel that is deficient in Ca is split open, often the small developing embryo will be brown instead of almost white. This symptom in the kernel has been termed black heart. Calcium-deficient kernels will not germinate properly, and those that do begin germinating may have malformed roots.

Sulfur

There is little need to apply sulfur (S) when fertilizing peanut. Large quantities of S are supplied when calcium sulfate is applied, and some S is available when ordinary superphosphate is applied. Sulfur has also been used extensively as the base for a fungicide. These sources, along with that from rainfall and irrigation water, seem to have prevented any S deficiency (Plate 13). Sulfur is retained because it is sorbed on clay surfaces, so there is usually an abundant supply in subsoils.

Micronutrients

Peanuts require the seven micronutrients known to be essential for plants: boron (B), chlorine (Cl), copper (Cu), iron (Fe), manganese (Mn), molybdenum (Mo), and zinc (Zn). Iron chlorosis is shown in Plate 14; Mn deficiency is shown in Plate 15; and Cu deficiency is shown in Plates 16 and 17. If the soil is quite acidic (pH 5.0), an application of Mo aids in N fixation and will increase yields. However, the recommended remedy for this situation is an application of lime to reduce acidity and provide extra Ca. The symptoms of this situation are those of N deficiency or a general chlorosis and reduced yields.

The micronutrient most often limiting for peanut production is B. It is likely to be deficient on coarse-textured, sandy soils such as used for this crop. Also, such soils are more subject to drought, and this condition enhances the need for additional B. Boron deficiency is expressed primarily in peanut quality rather than in yield.

Boron, like Ca, is required especially by the developing fruit, so it may be applied at any time until the fruiting period. However, because it is translocated better than Ca, it is available from the root system. It may be applied with other fertilizers or a fungicide or by any method that gets it on the field prior to fruiting. The rate required is only about 0.5 lb/acre (about 0.6 kilograms per hectare) of actual B. Rates greater than 1 lb/acre should not be used because toxicity symptoms may develop (Plate 18). This is especially true if the material is applied across the foliage. The symptoms of B toxicity are similar to those of salt accumulation. There is first a yellowing of the leaf margins. These margins will turn brown in time, and if the toxicity is severe, the symptoms will progress to the middle of the leaves and they will drop off.

Boron deficiency symptoms never occur in the tops of peanuts grown in the field. Symptoms develop in the fruit, especially in the inner face of the cotyledons. Often the cotyledon will not develop completely, leaving a depressed area in the center of the inner face. This area may become brown, and it is even more likely to become brown on roasting. This symptom has been termed hollow heart. If this problem is significant, the crop grade (Chapter 5) will be reduced. There is probably also some reduction in yield, but it has been too small to measure.

The peanut is quite sensitive to Zn, and the plant will show toxicity symptoms more readily than a number of other field crops. This condition occurs if the Mehlich-1 extractable soil Zn is greater than 12 milligrams per kilogram and/or the leaf Zn concentration is greater than 220 milligrams per kilogram and the soil pH is about 6. Uptake of Zn, however, is pH dependent, so if the soil is more acidic there will be more uptake, and hence more toxicity, at lower soil Zn levels.

Symptoms of Zn toxicity are reduced leaf size and stem length, giving the plant a dwarfed appearance (Plates 19–21). There may be some chlorosis of leaves and purpling of stems, but stunting is the most noticeable symptom and stem splitting near the base of the plant is the most diagnostic symptom. Under severe conditions, necrosis of the plant occurs quickly.

Summary

Fertilization of peanuts includes the following: the crop should be limed properly to keep pH between 6.0 and 7.0; it fixes its own N and grows well and produces good yields by using residual P and K applied to other crops in the rotation; it needs a little B to ensure quality; and it has a unique need for Ca, which seems to increase with the kernel size of the type grown. In general, the crop has probably been overfertilized in the past, but considering its high value, that is understandable.

Selected References

de Geus, J. G. 1973. Fertilizer Guide for the Tropics and Subtropics. 2nd ed. Centre d'Etude de l'Azote, Zurich.

Pattee, H. E., and Young, C. T., eds. 1982. Peanut Science and Technology. American Peanut Research and Education Society, Yoakum, TX.

Plucknett, D. L., and Sprague, H. B. 1989. Detecting Mineral Nutrient Deficiencies in Tropical and Temperate Crops. Tropical Agricultural Series, no. 7. Westview Press, Boulder, CO.

Power, J. F., ed. 1987. The Role of Legumes in Conservation Tillage Systems. Soil Conservation Society of America, Ankeny, IA.

Darold L. Ketring
U.S. Department of Agriculture
Agricultural Research Service
Stillwater, Oklahoma

Julia L. Reid
Department of Biological Science
Arkansas State University, State University

CHAPTER THREE

Peanut Growth and Development

Optimum conditions for growth and development of peanut are rarely met. An understanding of the conditions necessary for optimum growth provides a way to assess why expectations may not have been met. The health of a plant is not solely linked to the presence or absence of disease; many factors influence plant health (Chapter 1). Vigorous, healthy plants are generally more tolerant of biotic and abiotic stresses that may occur during the growing season. To begin the growing season with a healthy crop, high-quality seeds (greater than 80% germination) are necessary to provide good stands.

Planting Peanut Seed

Sowing high-quality seed in a well-prepared, moist seedbed (Chapter 2) is essential for crop establishment. Peanut seeds are generally planted at a depth of 1.5–2.0 inches (4–5 centimeters) at 80–110 pounds per acre in rows spaced at 3 feet. Seasonal variations in quality are common, even when the same standard practices of growing (nutrition and water), harvesting (digging and curing), and processing (storage, shelling, and fungicide treatment) are used. However, certain precautions can be taken to maintain the initial, inherent quality of the peanuts harvested for seed.

Germination and seedling emergence are under genetic control but are strongly influenced by environment. A cultivar with a history of a high percentage of germination and emergence, regardless of growing location or year, is a good choice to select for planting.

Adequate soil moisture at planting time is important for germination and seedling emergence. Low soil moisture can result in poor germination and emergence, leading to spotty, uneven stands. Regardless of the temperature, seed will not germinate without proper soil moisture. Generally, runner-type seed that ride a 17/64-inch-wide and Spanish-type seed that ride a 15/64-inch-wide or wider, 3/4-inch-long, slotted screen will provide the best emergence and seedling stands because these are the most mature seeds. However, smaller mature seeds (those that are firm and have tight, unwrinkled seed coats) also can provide good seedling stands. Under moderate to severe soil and/or climatic stress, small seeds may not initially perform as well as larger seeds, but final yields have not been found to be significantly affected by the size of seed.

After planting in moist soil, water uptake is the first phase in returning the dry seed to active growth. Water uptake is rapid during the first few hours and then proceeds slowly as the seedling grows. Minimum soil temperatures should be 60–65°F (16–19°C) for most of the day-night cycle. For more rapid emergence, soil temperatures above 70°F (21°C) are needed. The optimum temperature for the most rapid germination and seedling development is about 86°F (30°C). The radicle (embryonic root) will emerge from vigorous Spanish-type seed within 24 hours, but Virginia-type seed require 24–48 hours from the start of water uptake. Figure 3.1 illustrates the stages in peanut seed germination, seedling development, and emergence of the shoot above the soil surface. Initially, the radicle emerges from the seed (Fig. 3.1A) followed by rapid elongation of the taproot (Fig. 3.1B and C). Lateral roots begin to grow by 5 days after planting (Fig. 3.1D), and by 7 days, the shoot has emerged from the soil and many lateral roots have begun to grow (Fig. 3.1E). The time required for these stages to occur is dependent on temperature and seed quality. Longer times (10–14 days) may be required for seedling emergence at cool temperatures or with seed of low vigor. The growth rate of the hypocotyl (Figs. 3.1E and 3.2) determines how quickly the shoot will break through the soil surface and begin the manufacture of food from carbon dioxide, water, and soil nutrients using sunlight for energy. The taproot begins rapid growth before the hypocotyl and extends into the soil for uptake of water and nutrients (Fig. 3.2).

Crop Growth

Once the shoot emerges above the soil surface, the seedling is no longer dependent on the food reserves in the seed cotyledons. With sunlight for energy, carbon dioxide, water, and soil nutrients, the seedling is capable of manufacturing food for growth. Growth rate of the plant, however, is still highly dependent on temperature.

11

Temperature

Day-Degrees and Development. Temperature is a major environmental factor that determines the rate of peanut crop development. Temperatures above 95°F (35°C) inhibit the growth of presently grown peanut cultivars. The amount of temperature "heat" received by the crop can be measured in units called day-degrees (Box 3.1).

Developmental stages (Chapter 1) of peanut are highly correlated with day-degrees. A study in Oklahoma compared the progress of vegetative development (as indicated by the number of nodes added to the main axis) and reproductive (R) stages with day-degrees at two levels of irrigation. Emergence occurs at about 90 day-degrees. Flowering (stage R1) in the Spanish type begins at 300–400 day-degrees; more day-degrees are accumulated for flowering of the runner type. Under well-watered conditions of 33 inches (83 centimeters), there is a steady addition of nodes to the main axis and nearly linear development in R stage, but the development of the runner type is about one R stage slower that of the Spanish type. With about 16 inches (41 centimeters) of water, development is slower. Fewer main-stem nodes are produced by both Spanish and runner types, and R stage development is delayed with less water. This clearly shows the effect of limited water supply on the development of a peanut crop. Field temperatures of 95–110°F (35–43°C), even with an adequate water supply, inhibit peanut development.

Dry Weight Accumulation. Generally, the pattern of dry matter accumulation of a peanut crop is similar to that of most annual plant species. There is a slow period of early vegetative growth followed by rapid growth to a nearly linear maximum growth rate and then slow growth again toward the end of the season. Slowing of vegetative growth is associated with the beginning of reproductive growth (pod addition and seed filling) in the middle to latter part of the season.

In the example of the well-watered (33 inches [83 centimeters]) crop, the highest crop growth rate occurred during a 28-day period of about 500–900 day-degrees. The total crop growth rate was about 166 pounds per acre per day (lb/acre/day) for the runner type and 149 lb/acre/day for the Spanish type during this period. At about 900 day-degrees, substantial pod addition had begun and the crop had attained full pod development (R4) to beginning seed-fill (R5) stages. The pod growth rate was highest at about 900–1,300 day-degrees. The pod growth rate was about 64 lb/acre/day for the runner type and 56 lb/acre/day for the Spanish type. During this period, runner and Spanish types attained stages of about R5 and R6, respectively. The Spanish type began pod addition before the runner type and was about one R stage ahead of the runner type until near the end of the season when the dry weight of runner-type pods exceeded that of the Spanish type. The shelling percentages (dry weight of seeds divided by dry weight of pods) were about 71 and 75% for the runner and Spanish types, respectively, which indicate a high degree of maturity for a peanut crop.

In contrast, the crop with reduced water had the highest crop growth rate during a 30-day period of about 600–1,100 day-degrees. The total crop growth rate was about 104 lb/acre/day for the runner type and 107 lb/acre/day for the Spanish type during this period. Substantial pod addition was not evident until about 1,100 day-degrees for the Spanish type

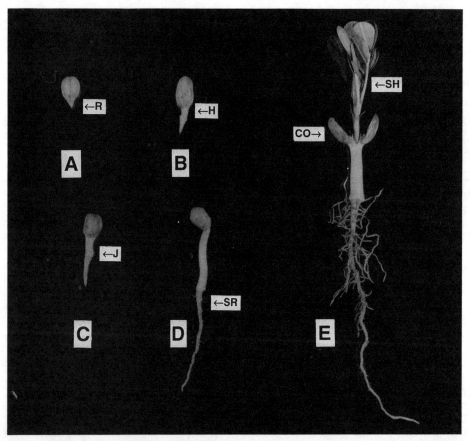

Fig. 3.1. Stages in peanut seedling development and emergence from the soil. **A,** 1 day after planting; R = radicle. **B,** 2 days after planting; H = hypocotyl. **C,** 3 days after planting; J = enlargement at the juncture of the radicle and hypocotyl. **D,** 5 days after planting; SR = lateral roots developing from the primary root. **E,** 7 days after planting; SH = primary shoot, and CO = cotyledons at the soil surface. (Courtesy H. Melouk and T. Johnson)

and even later for the runner type. The pod growth rate was highest during periods of about 900–1,300 day-degrees for the Spanish type and 1,100–1,430 day-degrees for the runner type. The pod growth rate was about 44 lb/acre/day (R6) for the Spanish type and 37 lb/acre/day (R5) for the runner type by the end of these periods. Pod dry weights were similar at the end of the season (in addition to the low amount of water, this season was shortened by an early frost). Shelling percentages were reduced to about 62 and 70% for the runner and Spanish types, respectively. These results indicate a major delay in crop growth and development caused by reduced water and high temperature.

Photosynthesis

The tetrafoliate leaf of the peanut plant houses the photosynthetic system (factory) for producing the basic metabolites (food) and energy-containing chemicals that are necessary for plant growth and development. Major products from photosynthesis are sugars, amino acids, organic acids, and energy-containing substances. These basic building blocks are used in the synthesis of complex carbohydrates, proteins, fats, nucleic acids, many small molecules (such as plant hormones) used in growth processes, and phytoalexins (inhibitory chemicals) used to fight plant diseases. In short, photosynthesis is the driving force for pod addition and seed filling.

Under field conditions, light is not a limiting factor of photosynthesis. Young leaves in the upper part of the canopy require nearly full sunlight (1,600–1,800 microeinsteins, i.e.,

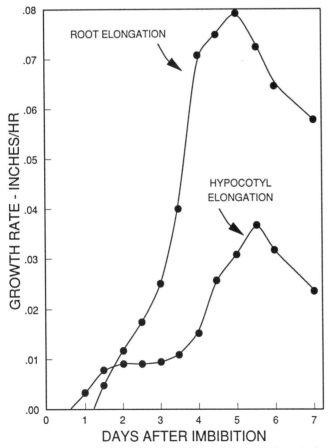

Fig. 3.2. Rates of elongation of the hypocotyl and main root of a medium-sized peanut (cultivar Starr) at 81°F (28°C). The decrease in elongation of the main root after 5 days is replaced by the initiation and growth of lateral roots. The decrease in hypocotyl elongation is likewise replaced by the growth of the epicotyl and leaves. (Reprinted, by permission, from Goeschl, 1975)

photons of light, per square meter per second) for light saturation. Canopy photosynthesis increases linearly nearly to full sunlight (Fig. 3.3). Leaves in the lowest part of the canopy, particularly after full ground cover, will begin to senesce and abscise (shed) from the plant because of the lack of light. Crop photosynthesis suffers from the same major limiting factors (temperature and water) as plant growth and development. Growth is very sensitive to small changes in internal plant water status. Thus, adverse environmental conditions and various pests (Chapters 8–10) affect photosynthesis, resulting in reduction of plant growth and reproduction (yield).

Photosynthesis attains its maximum rate at about 86°F (30°C), which is near optimum for vegetative growth. It decreases above this optimum so that under field conditions with temperatures of 95–110°F (35–42°C), photosynthesis is substantially reduced.

Water deficits, like excessive temperatures, also inhibit photosynthesis. Water deficits may occur in leaves because of high evaporative demand even though soil water content is high. Gradually, soil water content declines along with increases in temperatures. When these conditions persist over time, drastic reductions in yield can occur, particularly if drought and high temperatures coincide with sensitive growth stages.

Water deficits decrease photosynthesis by reducing leaf area (size of the factory), closing stomata (leaf pores that regulate the exchange of carbon dioxide and water vapor with the atmosphere), and decreasing the efficiency of the carbon dioxide fixation process (damage to machinery of the factory). Production of photosynthetically active leaf area is one of the most important factors affecting productivity and one of the most sensitive to water deficit. Both reduction in leaf size and production of fewer leaves can occur. Leaf pores begin to close, carbon dioxide uptake declines, and transpiration (water vapor release to the atmosphere) decreases when peanut leaf relative water content declines below about 85%. From 85% to about 30% relative leaf water content, photosynthesis and transpiration decline in a parallel manner. Under field conditions without additional water from irrigation or rainfall, soil moisture declines slowly from the initial water content (which is usually high because of spring rains prior to and after planting) to a low level at which roots are unable to extract any more water. In rain-fed agriculture (dryland), the crop

Box 3.1

Day-Degrees

The term "day-degrees" refers to thermal time, since the heat units are accumulated over time (days) during the growing season. Thermal time per day is estimated by dividing the sum of daily maximum and minimum temperatures by two and subtracting a base temperature, i.e., the temperature below which peanuts will not grow (56°F [13°C]). These daily values are totaled throughout the growing season. Developmental progress of peanut is linked to thermal time from planting to harvest. Calculated day-degree accumulation at 145 days after planting for three seasons, 1985, 1986, and 1987, in Oklahoma were 1,456, 1,672, and 1,473°C, respectively. Over the 3-year period, crop yields averaged about 2,750 pounds per acre for a Spanish-type and 3,185 pounds per acre for a runner-type genotype.

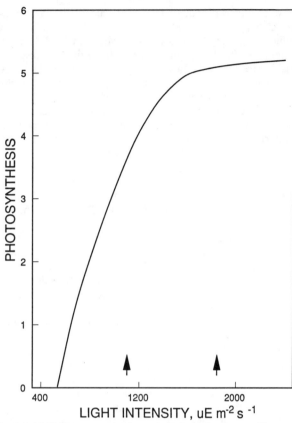

Fig. 3.3. Light response of apparent photosynthesis on a Florunner peanut canopy (light intensity measured in microeinsteins per square meter per second).

frequently is produced on stored soil water alone, but timely rains may occur in some years. When irrigation is available, supplemental water-use strategies need to be developed. It is beyond the scope of this chapter to discuss irrigation in detail, but some emerging concepts, which, it is hoped, will improve irrigation strategy for peanuts, are presented.

Water

Water is a simple molecule composed of two atoms of hydrogen and one of oxygen, but it is very complex in its behavior. Life, as we know it, cannot exist on Earth without water. There is much concern for water supplies of suitable quality for agriculture, industry, recreation, and human consumption. Groundwater reserves are declining, and maintenance of water of suitable quality is a serious problem. All of the processes mentioned above, from seed germination to production of the next seed generation (i.e., the crop life cycle), are dependent on water.

Soils. Soils provide the reservoir of water used by plants. The upper limit of water content held in the soil reservoir is the "field capacity" (the amount of water held in the soil after drainage caused by gravity), and the lower limit is the water content at which plants wilt and do not recover when rewatered. It is between these limits that plant roots function to extract water for growth and development. This amount of

Table 3.1. Diversity in root characteristics of peanut genotypes

Genotype and market type	Taproot length (inches)	Roots at 3-ft depth (number)	Root volume (oz. × 10)	Root dry weight (oz. × 10²)
Chico (Spanish)	60.6	1.2	7.0	6.0
Florunner (runner)	76.0	4.9	7.9	8.0
PI 355993 (Valencia)	63.8	3.0	10.9	9.0

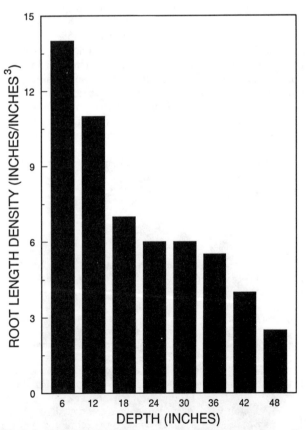

Fig. 3.4. Root length density of peanut plants at 80 days after planting in a Teller sandy loam soil. Plants were well irrigated.

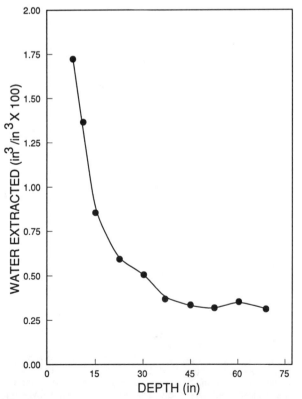

Fig. 3.5. Mean water extraction for 10 peanut genotypes in relation to depth in a Teller sandy loam soil.

water is termed "available water." But even this water is apparently not entirely extractable by roots. The amount of available water is dependent on soil texture. Sandy soils, which are typically used for peanut production, have less water-holding capacity, and hence less available water, than heavier loam and clay soils. However, plant roots have more difficulty penetrating heavier soils.

Roots. The extent of the soil water reservoir related to crop production is determined by the volume of soil penetrated by plant roots. The internal water balance of the plant changes in relation to the ability of roots to extract available water. There are genetic differences in the root and shoot characteristics of peanut plants and possibly in the ability of roots to extract water from soil under low soil water conditions. Genetic differences found in peanut plants are 1) depth of penetration of the taproot, 2) lateral root production, and 3) volume of the root system. Table 3.1 shows rooting differences in three diverse genotypes. In addition, peanut genotypes differ in their ability to extract water from a drying soil profile.

There are two strategies available to plants to extract the maximum available water from the soil profile: 1) more extensive root systems that penetrate deeply to explore more of the soil volume for water and 2) the ability to extract water to a greater extent as the soil dries, i.e., as the lower available water limit is approached. Thus, available water is a function of both the soil reservoir and the plant root system.

Determining the extent of roots in the soil profile can be accomplished by taking soil cores, cleaning the roots, and then measuring length and weight. This is a laborious, time-consuming process, but it provides essential information about root distribution in the soil profile. Root length densities

(inches of root per cubic inch of soil) of peanut plants determined by this procedure are shown in Figure 3.4. The highest root length density is in the top 6 inches (15 centimeters) of soil, and the least root length density is at the 48-inch (120-centimeter) depth.

Another, less laborious, means of estimating the location of roots in the soil profile is to determine the maximum depth at which water extraction can be detected with a soil water measuring device. Roots have been found to grow to a depth of about 110 inches (275 centimeters) in a deep, fine, sandy soil but only to 48 inches (120 centimeters) in a heavier sandy loam soil, as detected by water extraction (Fig. 3.5). Less water extraction was evident at 48 inches (120 centimeters) deep in this example soil, which agrees with the lower root length density found at this depth (Fig. 3.4). However, perhaps more important than water extracted at individual depths is the total water extracted by the entire root system from the soil reservoir.

Another factor affecting rooting density is plant spacing in the field. Dense plantings have lower amounts of roots relative to shoots, use water at faster rates, and thus deplete the soil profile of water sooner than sparse plantings. Sparse plantings, called skip-row (two rows planted and one fallow or two rows planted and two fallow), have higher yield and crop value per planted acre than solid plantings under drought conditions and in so-called average years, but there is no advantage under full irrigation. Denser plantings (19 versus five plants per square yard) or narrow rows (1 foot versus 3 feet) have increased pod yield, resulting in higher water-use efficiency, but disadvantages include the need for additional irrigations and, depending on cultivar, no increase in net crop value.

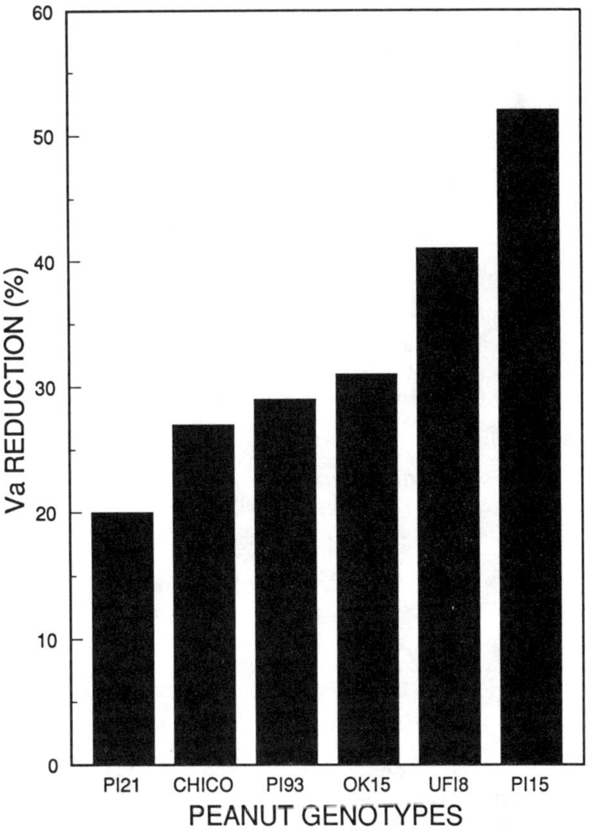

Fig. 3.6. Percentage of reduction in apparent sap velocity (Va) from irrigated to rain-fed conditions of six peanut genotypes: PI 404021 (PI21), Chico, PI 355993 (PI93), OK-FH15 (OK15), UF 77318 (UF18), and PI 405915 (PI15).

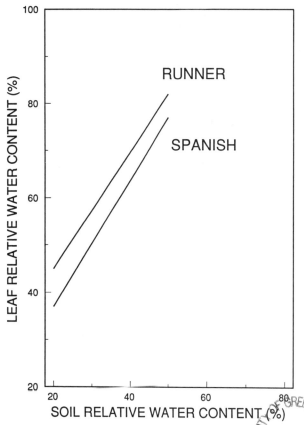

Fig. 3.7. Percentage of leaf relative water content in relation to percentage of soil relative water content for a runner-type (Florunner) and a Spanish-type (Comet) peanut cultivar.

Apparent sap velocity is a measure of water flowing in plants from roots to leaves. When comparisons were made among peanut genotypes for apparent velocity in a drying soil (rain-fed) versus an irrigated field, differences were found in their ability to maintain water flow to the shoots (Fig. 3.6). In other words, roots of some genotypes are more capable of extracting water from the drier soil, and as a consequence, there is less reduction in apparent velocity. This trait is very useful in utilizing more of the soil's available water.

The demand for water comes from the shoots (leaves) and is caused principally by air temperature, sunlight (which heats the leaves), and wind speed across the canopy. When soil water is plentiful (available water is high and soil relative water content is greater than 60%) (Fig. 3.7) and leaf relative water content is maintained equal to or greater than 85% (Fig. 3.7), then photosynthesis and growth proceed at optimum rate. As soil relative water content declines, leaf pores begin to close and leaf relative water content decreases to a level at which the plants begin to suffer from lack of water. Soil relative water content is available water expressed as a percentage. Thus, as soil relative water content approaches 45%, water becomes insufficient for optimum growth and development. In Teller sandy loam soil, soil relative water content decreases to this level about 55–65 days after planting, which is at the beginning of the pod addition phase of reproductive growth (Fig. 3.8). The pod addition and subsequent seed-filling stages are the most sensitive to lack of water. Insufficient water during these stages will cause the greatest reduction in yield. An excess of water during maturation toward the end of the season also can reduce yield.

Visual observation to estimate irrigation scheduling of peanuts can be misleading. Wilting of the plants on a hot, dry summer afternoon indicates the demand (leaves) for water exceeds supply capacity (roots), not necessarily that soil available water has reached the lower limit. If the plants do not recover overnight, insufficient soil water could be the problem, and an irrigation is necessary. These kinds of observations are not very useful, and more accurate means of irrigation scheduling are necessary for a high-dollar crop like peanut.

Water Application. Because of the complex interactions between the soil and plant water status, the atmospheric conditions that influence both of these, and the critical timing for water application, considerable research effort is being devoted to computer-assisted programs (models) to aid the grower in irrigation strategy. These models should also be beneficial in terms of economics, water conservation, and reduction of groundwater contamination. Peanut models have been developed by several scientists. The principal goal of the models is to improve our knowledge of the effects of environmental variables on peanut development, growth, pest invasion, water needs, and ultimately yield and quality. Another purpose for some of the models is irrigation management. Research and validation is in progress. The model from the National Peanut Laboratory is an expert system (computer aided) for management of irrigation and pest

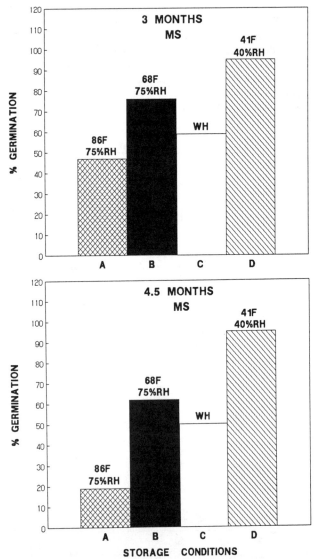

Fig. 3.8. Effect of storage environment (temperature and relative humidity [RH]) on germination of mechanically shelled (MS) peanut seeds after 3 and 4.5 months in storage. WH = warehouse conditions (68°F, 75% RH). (Adapted from Baskin and Delouche, 1971)

Box 3.2

Water Requirements

The desire is to provide optimum amounts of water from emergence to maturity. The total seasonal water necessary for peanut production is 16–30 inches (40–80 centimeters). The upper and lower limits vary annually, depending on daily weather (air temperature, soil temperature, rainfall, wind, and radiation). Some or all of these factors are used by computer models to estimate daily water use by the crop, and this amount of water is subtracted from the soil reservoir (i.e., available water). When available water reaches a prescribed level (for instance, 50%), the crop is rewatered. The extent that available water can be allowed to decline is dependent on the growth stage. Early season vegetative growth up to 50 days after planting is least sensitive to water deficit, and mild stress during this period has the least effect on final yields. Midseason water deficits during the pod addition and early seed-filling stages have the most inhibitory effect on yield and quality. Midseason also is the time of maximum water use by the crop. However, excess water during maturation phases can reduce yield and quality. Thus, the current strategy is to irrigate sparingly up to 50 days after planting, maintain available water level greater than 50% during midseason, and reduce irrigation again during late-season maturation. This is very difficult to accomplish, but with the concerns for water and crop management and the new modeling technology, it may be possible in the near future.

environment. The University of Florida and North Carolina State University models require daily input of weather data. Some of the data, such as solar radiation and wet bulb temperature, are not usually recorded without special provisions by researchers at agricultural experiment stations. Deployment of these models may first occur at these stations for use on a regional basis and eventually could provide general advisories for irrigation and disease forecasting. The expert system from the National Peanut Laboratory uses soil temperatures in the pod zone and water (irrigation plus rainfall) as the main inputs and has been successfully deployed for validation with extension agents and peanut growers. It provides irrigation and pest advisories and gives a prediction of potential yield, as do the other models.

Box 3.2 summarizes the water requirements of peanut and current irrigation strategies.

Harvesting and Storage of Seed Peanuts

Peanuts should not be dug if freezing temperatures are predicted for the time during which peanuts will be drying in

the windrow. Freeze damage of high-moisture peanuts reduces germination, seedling vigor, and metabolic processes of the seed. Even chilling temperatures that remain slightly above freezing will harm high-moisture peanut seeds.

Mechanical damage to pods and seed can occur from impacts during digging, combining, and shelling. Damage of this kind can be minimal if equipment is properly calibrated and operated. The protruding tip (sharp end) of the seed, which contains the taproot, is easily damaged during harvesting and handling. Another part of the seed that is easily damaged is the point at which the cotyledons (embryonic leaves) attach to the embryo. The large size of the cotyledons relative to the point of attachment to the embryo makes this region susceptible to breakage, particularly if too rapid drying has loosened the seed coats. Breakage of this attachment prevents food reserves in the cotyledons from being available to the embryo during germination and initial stages of seedling growth.

Mechanically shelled seed undergo rapid decrease in germinability under adverse storage environments. With few exceptions, the rate of decrease in germinability is proportional to the relative humidity (RH) and temperature. For

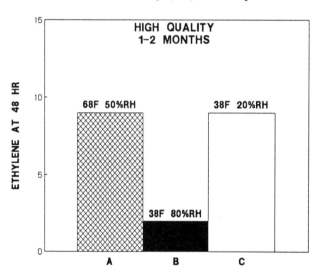

Fig. 3.9. Effect of storage environment (temperature and relative humidity [RH]) on germination of high-quality (80% germination) and low-quality (30% germination) seed lots. (Adapted from Ketring, 1971)

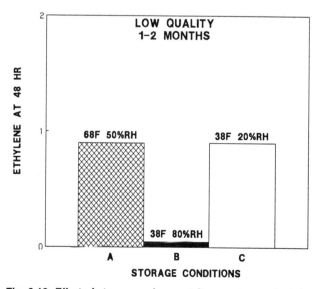

Fig. 3.10. Effect of storage environment (temperature and relative humidity [RH]) on ethylene production by high- and low-quality seed lots. (Adapted from Ketring, 1971)

example, a temperature of 86°F (30°C) and RH of 75% has been shown to cause more rapid decline in germination of mechanically shelled seed than a lower temperature and RH (Fig. 3.8). Warehouse conditions of about 68°F (20°C) and 75% RH also are detrimental to germination. Seeds retained over 90% germination when stored at 41°F (5°C) and 40% RH. Initial germination of high-quality seed (greater than 80% germination) is affected less by poor storage conditions than that of low-quality seed (less than 50% germination). All seeds are susceptible to rapid deterioration at high RH (80%), even in cold storage at 38°F (3°C) (Fig. 3.9). Ethylene, a plant hormone that regulates peanut seed germination, is also affected by poor storage conditions. High-quality seed produces about 10 times more ethylene than low-quality seed (Fig. 3.10). Ethylene production and germination are both reduced by high RH, even in cold storage; damage occurs to a greater extent in low-quality seed (Fig. 3.10). The adverse effect of high RH is enhanced by temperatures above 70°F (21°C). Seed quality can be adequately maintained in cold storage at 40°F (4°C) and low RH (50–60%), but good-quality seed will remain viable and vigorous for up to 1 year when stored at room temperature (68–75°F [20–24°C]), provided the RH is kept low. As a rule, with temperatures of 35–45°F (2–8°C) and RH of 55–65%, the total of RH plus temperature for acceptable peanut storage is about 100. This presents a major concern for the commercial peanut seed industry because the bulk of peanut pods and seeds necessitates large storage facilities, which are subject to fluctuations in ambient temperatures and RH. However, storage in these large facilities has been reasonably successful because the major storage period is during the cool winter months.

Most peanuts used for seed are removed from storage and shelled just before sowing. The seeds are usually treated immediately after shelling with an approved chemical (Chapter 11). Peanuts shelled several weeks before sowing and not chemically treated should be placed in cold storage. However, even with these precautions, problems continue to arise with peanut seeds that do not germinate and produce a seedling after planting, which could be the result of periods of high temperature and RH during storage. On the whole, a grower buying peanut seeds can depend on a reliable commercial seedsman to provide good quality, certified seed. The grower has the responsibility of protecting the seed from mechanical damage and high temperature and RH during the time from purchase to planting.

Selected References

Baskin, C. C., and Delouche, J. C. 1971. Effects of mechanical shelling on storability of peanut (Arachis hypogaea L.) seed. Proc. Assoc. Off. Seed Anal. 61:78-84.

Boote, K. J. 1982. Growth stages of peanut (Arachis hypogaea L.). Peanut Sci. 9:35-39.

Boote, K. J., and Ketring, D. L. 1990. Peanut. Pages 675-717 in: Irrigation of Agricultural Crops. B. A. Stewart and D. R. Nielsen, eds. Monogr. 30. American Society of Agronomy, Madison, WI.

Clark, L. E. 1975. Seed quality. Pages 10-18 in: Peanut Production in Texas. Texas Agricultural Experiment Station and Texas Agricultural Extension Service, College Station.

Goeschl, J. D. 1975. Germination and emergence. Pages 34-37 in: Peanut Production in Texas. Texas Agricultural Experiment Station and Texas Agricultural Extension Service, College Station.

Ketring, D. L. 1971. Physiology of oil seeds. III. Response of initially high and low germinating Spanish-type peanut seeds to three storage environments. Agron. J. 63:435-438.

Ketring, D. L., Brown, R. H., Sullivan, G. A., and Johnson, B. B. 1982. Growth physiology. Pages 411-457 in: Peanut Science and Technology. H. E. Pattee and C. T. Young, eds. American Peanut Research and Education Society, Yoakum, TX.

Ketring, D. L., Simpson, C. E., and Smith, O. D. 1978. Physiology of oil seeds. VII. Growing season and location effects on seedling vigor and ethylene production by seeds of three peanut cultivars. Crop Sci. 17:409-413.

Ketring, D. L., and Wheless, T. G. 1989. Thermal time requirements for phenological development of peanut. Agron. J. 81:910-917.

Wynne, J. C., and Sullivan, G. A. 1978. Effect of environment and cultivar on peanut seedling emergence. Peanut Sci. 5:109-111.

CHAPTER FOUR

O. D. Smith
Department of Soil and Crop Science
Texas A&M University, College Station

C. E. Simpson
Texas Agricultural Experiment Station
Texas A&M University, Stephenville

Selection of Peanut Cultivars

Cultivar selection is one of the most important decisions in successful peanut production. The seed holds the genetically determined yield potential of the crop as well as the ability of the crop to approach that potential when plant health is less than ideal. A last-minute decision based upon whatever the local seedsman has on hand is inadequate. Time is well spent seeking reliable and unbiased information regarding the strengths and weaknesses of available cultivars and evaluating how well their strengths fulfill the most important needs of each grower's specific production and marketing situation.

Objective evaluation of many field tests within the area under consideration for planting is important. Cultivar choice based on results from distant tests, such as from other states, or on conditions very different from those anticipated should be made with great caution. Yields of cultivars vary from test to test and from year to year within local areas because of varying plant health, which depends on soil, pest, weather, and cultural factors. Often, small differences in growing conditions cannot be fully explained or anticipated; thus, multitest averages are the best predictors of future performance, unless a change in conditions can be anticipated.

Market Type

Selecting a market type is often the first decision when considering cultivars. Market-type designations are sometimes confused with cultivar names. Type is principally a market designation based on seed size, although there are some botanical differences between the Spanish and Valencia types compared with the Virginia and runner types of peanuts (Chapter 1, Table 1.2). There are several cultivars within each market type.

Spanish and Valencia peanuts belong to the subspecies *fastigiata*. This group of peanuts is characterized by upright growth habit, flowers on the main stem, pegs at consecutive leaf attachment positions (nodes) on lateral branches, and limited or no fresh seed dormancy. Spanish peanuts typically have two seeds per pod, smooth seed coats, and a higher oil content than other market types. Valencia peanuts have high proportions of three or more seeds per pod and taste sweeter than other types. The currently marketed U.S. Valencia cultivars have red seed coats (testae).

Runner and Virginia are market class subdivisions of the subspecies *hypogaea* and are similar in most respects except pod and seed size. In this botanical group, fruiting on the lateral branches is at alternate nodes, and fresh seed dormancy is common. Runner and Virginia peanuts may be decumbent (spreading) or have a bunching growth habit. The principal factor determining whether a cultivar is marketed as a runner or a Virginia peanut is pod size. Cultivars of this subspecies with more than 40% so-called fancy pods (those with a minimum diameter greater than 34/64 inch [1.3 centimeters]) are marketed as Virginia; those with less than 40% are marketed as runner. The more desirable Virginia-type cultivars also have larger seeds, that is, seeds that are retained on, or ride, a 21.5/64- × 1.0-inch (0.85- × 2.54-centimeter) slotted screen, and premiums are paid for these large seeds.

Traditionally, peanuts planted in the Virginia and North Carolina area have been the large-pod Virginia cultivars, and use of these cultivars continues today. Runner types were grown in the Southeast (Alabama, Georgia, and Florida), Spanish in Oklahoma and Texas, and Valencia in New Mexico. As a result of the highly successful cultivar Florunner, however, more than 95% of the Southeast and 50% of the Southwest production is now runner. From a production standpoint, all market types could be grown in any of the major peanut-production areas, except perhaps where the growing season is limited because of lack of rainfall or a short duration of warm temperatures. However, the diversity in peanut pod size requires different storage and processing procedures. Thus, the use of multiple market types within a production area may strain the marketing system and make production of the unusual market type impractical for the industry.

The best assurance of getting seed of the cultivars desired is through the purchase of certified seed. Seed certification is managed under the authorization of legally recognized state agencies, which are usually land grant institutions or the state departments of agriculture. Four designations of seed are produced: breeder's, foundation, registered, and certified. Breeder's seed is produced under the direction of the peanut breeder and is used for the production of foundation seed. Foundation seed, produced under the auspices of foundation seed organizations, is made available to growers with preference in some states to certified seed growers. Registered and certified classes of seed are progressive increases from foundation seed and must be produced, processed, and marketed in accordance with prescribed guidelines aimed at ensuring variety purity and high seed quality. The intended use of certified seed is planting for commercial production. Both registered and certified seed are sold by the seed producers. A list of these growers should be available from the certification agency in each state and the state agricultural extension service. Periodic inspections and analyses are required during the production and processing of certified seed to assure purchasers that the seed is the cultivar specified and is not contaminated with noxious weed seed.

Yield Potential

In general, the Virginia botanical group of peanut, represented by the Virginia and runner market types, has the potential to produce higher yields than the Spanish and Valencia types. The Virginia group, as a whole, requires both a longer growing season and more moisture on a daily basis during the growth phase than the Spanish-Valencia group. The longer growing season might account for a portion of its higher yield potential. The phenomenon of late-maturing cultivars having higher yield potential is recognized in many crops. As a result of this high yield potential, the late-maturing Virginia peanuts often produce more total tonnage than the early maturing Spanish peanuts. If digging is premature (i.e., before 75–80% of pods are mature), immature fruit can result in processing and utilization characteristics that could negatively affect the marketability of the crop.

Growing Season

The length of the growing season required for the several types of peanut varies widely, but it takes Virginia and runner peanuts, in general, 1–6 weeks longer to mature than Valencia and Spanish peanuts. Late-maturing Virginia and runner peanuts attain full maturity in 130–150 days in the eastern United States but may require more than 170 days in Oklahoma and northwest Texas. The basis for this apparent disparity is not fully understood; but temperature, relative humidity, and elevation are likely factors. Within market types, the range in times to maturity among cultivars is generally less.

Stand Establishment

Seedling vigor, a factor in stand establishment and an important component of plant health, results from environmental and genetic factors. The importance of good seedling vigor is most notable when growing conditions impose stress. For example, Florunner tends to emerge slowly compared with most other runner cultivars and, when planted in cool soil, may require 14–25 days for emergence. Cultivars such as Tamrun 88 and GK-7 may emerge 5–10 days sooner. This delay in emergence is not usually a serious production factor unless rainfall or seedling disease induces additional stress.

Seed dormancy is another factor that may affect stand establishment. Most Virginia and runner cultivars have moderate preharvest and postharvest seed dormancy; Spanish and Valencia peanuts typically have little dormancy. Seed dormancy is beneficial for production because it prevents or minimizes preharvest sprouting if digging is delayed beyond maturity. However, seed dormancy may persist for several months in Virginia and runner cultivars, depending upon harvest and storage conditions. This can be a factor in cultivar selection if early spring plantings or plantings for two generations per year are contemplated, since poor, erratic, or delayed stands may result from unbroken dormancy. No commercial treatment has been approved for breaking seed dormancy.

Soil Factors

Although the peanut plant will grow on most soils if the appropriate nutrients and moisture are available and the appropriate rhizobium bacteria are present (Chapter 2), sandy soils are ideal for effective mechanized digging of peanut. Peanut cultivars with weak peg attachment or dispersed pod distribution are normally susceptible to pod loss at digging. On the whole, Spanish peanut cultivars sustain less mechanical pod loss as a result of soil texture and compaction than runner and Virginia types. There are fewer differences among cultivars of the same market type than among cultivars of different market types. Peg strain during digging is increased by pod size; thus, large-pod cultivars (e.g., NC 7) are more suitable for very sandy soils that have little crusting. The combination of large pods, weak pegs, and heavy soil greatly increases the danger of pod loss.

Nutrients that affect cultivar selection include calcium and boron (Chapter 2). Large-pod cultivars require more calcium and boron in the pod-development soil zone than other types for good pod production and seed fill.

Resistance

Disease

Options for curtailing disease problems by cultivar selection are limited but may become more important in the future. All disease resistance reported in U.S. cultivars is of a partial nature; that is, all cultivars can be infected, but some cultivars have more disease resistance than others. A summary of cultivar characteristics, including known disease and pest resistance, is shown in Table 4.1. Resistance to late leaf spot, rust, and web blotch (Chapter 10) was an important factor in the release of Southern Runner. Subsequent research has shown this cultivar to have some resistance both to tomato spotted wilt virus (Chapter 9) and to stem rot (southern blight or white mold) caused by *Sclerotium rolfsii* Sacc. However, Southern Runner and Langley have been especially susceptible to Sclerotinia blight at Stephenville, Texas. Some evidence of partial resistance to early leaf spot has been noted in Florunner, which has also shown some resistance to web blotch. Toalson was released because of its moderate resistance to Pythium and Rhizoctonia pod rot. Subsequent research has shown it also has good resistance to Sclerotinia blight caused by *Sclerotinia minor* Jagger and some resistance to stem rot. Justification for the release of Tamspan 90 included resistance to Sclerotinia blight and pod rot, the resistance supposedly inherited from its Toalson parentage. However, it has not shown resistance to stem rot. The cultivar Southwest Runner was released in 1995 because of its resistance to Sclerotinia blight. The cultivars NC 8C and NC 10C were released because of resistance to Cylindrocladium black rot, and Va 81B and AD 1 have some resistance to Sclerotinia blight.

Generalizations are often dangerous and certainly are not without exception, but a few observations regarding market types as represented by the current cultivars can be made: 1) the Spanish cultivars are usually less susceptible to the important pod-rotting organisms than the runner and Virginia cultivars; 2) the Spanish cultivars are generally less susceptible to Sclerotinia blight than the runner and Virginia cultivars; and 3) the Virginia and runner cultivars tend to show less early season injury from leaf spot than the Spanish cultivars.

Insects

Few reports have been made of resistance to insects in currently grown peanut cultivars. NC 6 is resistant to southern corn rootworm and leafhoppers (Chapter 8). Leafhopper injury symptoms on peanuts are often more prominent on certain runner cultivars than on Spanish cultivars; however, this difference might be amplified by the contrast in color between the dark green runner and less intense green Spanish cultivars, which makes the symptoms more apparent. Lesser cornstalk borer damage was reported as being less in Florunner than in

Starr, but Florunner losses can be severe under dryland culture when conditions are right.

Nematodes

No nematode resistance among current cultivars is known, but nematodes are an important pest in several production areas (Chapter 12). Transfer of nematode resistance from wild species is in progress.

Quality

"Quality" is a loosely used term that has varied meanings among the segments of the industry. Growers speak of quality in terms of grade, basically implying that a cultivar with high quality has a high percentage of sound, mature kernels or produces pods with bright color (if it is to be sold for the in-shell trade). These are the quality factors that impact the producers directly.

Seedsmen and growers refer to seed quality with regard to uniformity, seed size, and seedling vigor. Shellers and brokers consider foreign material, contaminants, and uniformity in

Table 4.1. Peanut cultivar characteristics

Market type Cultivar name	Releasing state agency or company	Maturity group[a]	Vine type[b]	Leaf color[c]	Pod size[d]	Seed size[d]	Reported pest resistance[e]	Greater pest susceptibility
Spanish								
Pronto	Oklahoma, USDA[f]	1	U	LYG	S	S	None	
Spanco	Oklahoma, USDA	1	U	LYG	S	S	None	
Starr	Texas	1	U	LYG	S	S	None	
Tamnut 74	Texas, Oklahoma, Georgia	2	U	LYG	S	S	None	
Tamspan 90	Texas, USDA	1	U	LYG	S	S	*Pythium, Sclerotinia*	
Toalson	Texas	2	U	LYG	S	S	*Pythium, Rhizoctonia, Sclerotinia*	
Valencia								
Georgia Red	Georgia	2	U	LYG	M	M	None	
NM Val A	New Mexico	2	U	LYG	M	M	None	
NM Val C	New Mexico	2	U	LYG	M	M	None	
Runner								
AT 127	AgriTech	3	S	DG	L	L	None	
Florunner	Florida	4	S	DG	L	L	None	
GK-7	AgriTech	4	S	DG	L	L	None	
Langley	Texas	3	S	DG	L	L	None	*Sclerotinia*
MARC-1	Florida	3	S	DG	L	L	None	*Sclerotinia*
Okrun	Oklahoma, USDA	4	S	DG	L	L	Pod rot	
Southern Runner	Florida	5	S	G	L	L	Late leaf spot, rust, web blotch, TSWV, stem rot	*Sclerotinia*
Sunrunner	Florida	4	S	DG	L	L	None	
Tamrun 88	Texas	4	S	DG	L	L	None	TSWV
Southwest Runner	Oklahoma, USDA	4	S	DG	L	L	*Sclerotinia*	
Georgia Browne	Georgia	5	S	DG	S	S	Stem rot, limb rot, TSWV	
Virginia								
AD 1	Keel Peanut Co.	5	B	DG	XL	L	*Sclerotinia*	
Florigiant	North Carolina, Florida	5	S	DG	XL	XL	None	
NC 6	North Carolina	5	S	DG	XL	L	Leafhoppers, SCR	
NC 7	North Carolina	4	B	DG	XL	XL	None	
NC 8C	North Carolina	5	B	DG	XL	L	CBR, SCR	
NC 9	North Carolina	4	B	DG	XL	L		
NC 10C	North Carolina	5	B	DG	XL	XL	CBR	
NCV-11	North Carolina, Virginia	4	S	DG	L	L	None	
VA 81B	Virginia, USDA	4	B	DG	XL	L	*Sclerotinia*	

[a] 1 = 110–130 days; 2 = 120–135 days; 3 = 126–140 days; 4 = 130–145 days; and 5 = 140–165 days.
[b] U = upright; S = spreading; and B = bunch.
[c] LYG = light yellow green; DG = dark green; and G = green
[d] S = small; M = medium; L = large; and XL = extra large.
[e] TSWV = tomato spotted wilt virus; CBR = Cylindrocladium black rot; and SCR = southern corn rootworm.
[f] U.S. Department of Agriculture.

assessing quality. Manufacturers consider quality in terms of chemical composition, contamination, flavor, and storability. All these characteristics might be influenced by cultivar, but other factors are also involved.

In general, grades of U.S. cultivars tend to be higher for mature, properly managed runner peanuts than for the other market types. Pod brightness is affected by pod disease, smoothness of the pod surface, and the presence of microscopic hairlike structures on the pod surface to which soil tends to adhere. Cultivars with small pods and seeds are more difficult to clean of foreign material.

Oil composition is an important factor in shelf storage and the onset of rancidity. The oleic:linoleic acid (O/L) ratio and iodine value of peanut oil (dry peanut seed is 44–56% oil) are often used as indicators of shelf life. Peanuts with high O/L ratios (greater than 2.0:1) and low iodine values (less than 90 from a range of 82–107) are preferred. The O/L ratios of runner and Virginia peanuts are frequently above 2.0:1, while those of Spanish peanuts are often less than 1.5:1. Cultivar differences in oil composition, flavor, and other factors that affect product quality are apparent, and greater emphasis on these traits is expected in the future as marketing becomes increasingly more competitive. Although the direct effect of product quality factors on growers is small at present, such characters impact the industry and might become increasingly important in the future.

Fortunately, several good cultivars are available in each of the peanut-growing areas of the United States, and thus several good choices can often be made. Since many of the production costs will be common to all cultivars, the cultivar that can best fulfill expectations of the local production system in withstanding the production constraints common to the area will usually be the cultivar of choice. Local, regional, national, and international market demands can also have a major influence on the choice of both market type and cultivar.

Selected References

Ahmed, E. A., and Young, C. T. 1982. Composition, quality, and flavor of peanuts. Pages 655-688 in: Peanut Science and Technology. H. E. Pattee and C. T. Young, eds. American Peanut Research and Education Society, Yoakum, TX.

Cobb, W. Y., and Johnson, B. R. 1973. Physicochemical properties of peanuts. Pages 209-263 in: Peanut Culture and Uses. American Peanut Research and Education Association, Oklahoma State University, Stillwater.

Gregory, W. C., Smith, B. W., and Yarbrough, J. A. 1951. Morphology, genetics and breeding. Pages 28-88 in: The Peanut: The Unpredictable Legume. The National Fertilizer Association, Washington, DC.

Knauft, D. A., and Gorbet, D. W. 1989. Genetic diversity among peanut cultivars. Crop Sci. 29:1417-1422.

Knauft, D. A., Norden, A. J., and Gorbet, D. W. 1987. Peanut. Pages 346-384 in: Principles of Cultivar Development. Vol. 2, Crop Species. W. R. Fehr, ed. Macmillan, New York.

Mozingo, R. W., Coffelt, T. A., and Wynne, J. C. 1987. Characteristics of Virginia-type peanut varieties released from 1944–1985. South. Coop. Ser. Bull. 326.

Norden, A. J., Smith, O. D., and Gorbet, D. W. 1982. Breeding of the cultivated peanut. Pages 95-122 in: Peanut Science and Technology. H. E. Pattee and C. T. Young, eds. American Peanut Research and Education Society, Yoakum, TX.

Smith, O. D., Simpson, C. E., Harrison, A. L., and Spears, B. R. 1975. Variety characteristics. Pages 1-5 in: Peanut Production in Texas. Tex. Agric. Exp. Stn. Publ. RM-3.

Wynne, L. C., and Gregory, W. C. 1981. Peanut breeding. Adv. Agron. 34:39-72.

Timothy H. Sanders

U.S. Department of Agriculture
Agricultural Research Service
Raleigh, North Carolina

CHAPTER FIVE

Harvesting, Storage, and Quality of Peanuts

In the United States, the driving force behind sales of peanuts is the unique, high-impact flavor of the roasted product. Flavor can thus be considered a major measure of quality. A simple definition of quality is "consumer acceptability," and since there are many consumers from the farm to the pantry shelf, there are many measures of quality. All factors that make a peanut excel in flavor, appearance, weight, and biochemical makeup contribute to its quality. Quality factors are related and interactive. For example, if there is plenty of water and the temperature is warm early in the season, a grower might be heard saying, "Looks like we will have a high-quality crop." His definition of quality may be that the plants are about the right size, flowering is progressing, pods are beginning to form, and the weather looks favorable. This measure of quality is closely related to all the measures of quality that come later. Quality is a focal point of the peanut industry. To obtain the highest measure of consumer acceptability is the goal of each seller. Research and extension workers have cooperated to arrive at many methods to aid in this task.

Production, harvesting, curing, storage, shelling, transporting, processing, and all of the associated steps add up to affect peanut quality. Interrelated quality factors are of concern from harvest through storage of shelled peanuts. In our discussion of quality, we must make the basic assumption that conditions up to harvest have been in the range considered adequate for acceptable peanut production. The effect of deviations from that range will of necessity be considered.

The quality of a load of farmer's stock peanuts is measured by the grade. Grading of peanuts usually involves determination of percent moisture and the percentages of sound, mature kernels (SMK), loose-shelled kernels (LSK), fancy pods, extra large kernels (ELK), damage, and foreign material. The grade of farmer's stock peanuts is a major determinant of price within the price support structure.

Harvesting

Harvest for Quality

The indeterminate flowering pattern of the peanut plant makes proper timing of harvest difficult, even though such timing is crucial for obtaining maximum yield, grade, and quality. Harvest at the proper time ensures that a high percentage of mature pods remain on the plants, that the maximum number of pods have attained their greatest weight, and that the most mature pods are not falling off and thus beginning a rapid decrease in harvestable yield. Optimum harvest may be complicated by the presence of many immature pods. This can be caused by a "split crop" situation, which results from a break in the flowering or pod set pattern. Delays in maturation may also occur because of late-season drought stress. Therefore, timing of peanut harvest is critical to many of the measures of peanut quality.

> **All factors that make a peanut excel in flavor, appearance, weight, and biochemical makeup contribute to its quality.**

Yield and grade are greatly affected by the timing of the harvest. These two factors determine the value received for the peanut crop by the producer. Timing of harvest affects seed size distribution, an important factor in determining the value of peanuts after shelling. Harvest timing also determines the maturity composition of peanuts in final shelled lots. Timing relates not only to maturity but also to environmental conditions during the harvesting process. Long periods of rain immediately prior to harvest may result in both yield loss and deterioration of peanut quality. Extreme high temperatures while the crop is in windrows can cause too rapid drying and may contribute to development of off-flavors. Exposure to freezing temperatures for only a few hours results in off-flavors and reductions in grade. Timing of actual picking of the pods and equipment adjustments influence the amount of damage to pods and the amount of foreign material in harvested loads.

Harvest for Optimum Maturity

When the harvested crop contains an excessive number of immature pods, the potential for reduced flavor, storage, and shelf life quality is increased. On the other hand, if harvest is delayed, the largest, most mature pods fall off during picking, reducing yield and dollar return (Table 5.1). Delayed harvest may actually result in a crop similar in composition to one that has been harvested too early; that is, there may be fewer mature pods in the load. Because many peanuts tend to enlarge before they mature, the potential reduced quality is not limited exclusively to the small seed in the lot. Immature peanuts are physiologically different from mature peanuts, and thus they respond differently to any particular set of conditions. The biochemical components of immature peanuts are not as stable as those of mature peanuts. Exposure to severe conditions, such as freezing or high temperature, results in greater quality deterioration with immature peanuts than with mature ones. Plate 22 illustrates the maturity composition of a crop of peanuts near harvest.

Immature peanuts do not have as much potential for full flavor development as mature peanuts, and they have greater potential for off-flavor development, even under very gentle curing conditions. Curing studies have demonstrated that the flavor of immature peanuts is negatively affected when they are cured at temperatures only 15°F above ambient conditions (ambient maximum of 92°F [33°C]). As curing temperatures are increased to 30°F above ambient, progressively more maturity classes are affected. The overall flavor of shelled, sized lots is thus a combination of the flavor of the different maturity classes contained in the lot. The greater the percentage of immature peanuts in any lot, the greater the potential for flavor and quality reduction.

A harvested lot with a high percentage of immature pods may also have a high moisture content and will be difficult to dry to a uniformly safe storage level. Uneven moisture distribution may result in overdrying of mature pods and an increase in the percentage of split seed. When mature pods are cured to optimum moisture content, immature pods may still range to 18% moisture or higher (after curing). High-moisture peanuts in storage have a high probability of deterioration in quality and an increased potential for growth of *Aspergillus flavus* Link:Fr. and aflatoxin production.

Maturity Testing Methods

Methods to determine the proper time to dig the peanut crop have been used for many years and range from digging a certain number of days after planting to using highly sophisticated techniques that require expensive equipment. Under a given set of conditions, probably all the methods have some utility; however, some methods are so environmentally dependent or expensive that general use is precluded. It should be stressed that maturity testing methods are only tools. They must be used in conjunction with frequent field observations, since there are conditions of disease, weather, labor, and equipment management that can override any indicators of the optimum time of harvest. Harvest

of large acreages cannot be done quickly, and thus evaluating individual fields and understanding that some areas may need to be harvested early are essential in obtaining maximum yield. Generally, harvesting 1 week early results in less loss than harvesting 1 week late (Table 5.1). Presently, only two methods are widely used for peanut maturity determinations: the shellout method and the hull-scrape method.

Shellout Method. The shellout method is based on color changes within the hull that occur as peanuts mature. The internal hull surface in most varieties gradually changes from solid white to brown or black splotches that cover a high percentage of the area. The color of the seed coat changes from white to dark pink or light tan at the same time. Maturity determination with this method uses all but the most immature (soft, watery) pods from several plants from representative locations in a field. The most immature pods are not utilized because they will not be harvested (because of their light weight, they will be blown out of the combine with the vines). The pods are opened, and the percentage of pods with tan to brown color inside the hull and pink to dark pink seed coats is determined. Some variability exists in the recommended percentage at which harvest should begin, although the range is generally 60–80%, depending on cultivar and environmental factors. The suggested percentages for harvest of the three major market types are runner, 70–80%; Virginia, 60–65%; and Spanish, 75–80%. Valencia market types should be treated in a manner similar to that of Spanish types since neither has fresh seed dormancy. Because of the lack of fresh seed dormancy, sprouting can be a problem if Valencia and Spanish types are allowed to become overmature.

When the recommended percentages of mature seed cannot be achieved because of irregular pod set, drought, or other factors, judgment is required. The use of any percentage should be examined in relation to plant condition. Disease-free plants

Table 5.1. Average yield and dollars lost over 4 years from digging Florunner peanuts too early or too late

Relative week	Yield lost (lb/acre)	SMK + SS[a] (%)	Value per ton[b] ($)	Value lost per acre[b] ($)
−2	778	73.9	654	254
−1	222	74.2	657	73
0	0	75.0	663	0
+1	556	75.6	668	186

[a] SMK = sound, mature kernels (whole), and SS = sound, split kernels (kernels that have been split in half).
[b] Based on 1989 USDA quota support price.

Box 5.1

Shellout Maturity Testing Method

- Select five to 10 plants from representative areas of the field.
- Pick off all combine-harvestable pods (i.e., soft, watery pods that shrivel in windrows should not be used).
- Crack open each pod to examine internal hull and seed coat color.
- Place pods with tan to black internal hull color and pink to dark pink seed coat color together as mature pods.
- Calculate the percentage of mature pods:

  % mature pods = [number of mature pods/(number of mature pods + number of immature pods)] × 100

- Mature pod percentages for approximate harvest time: runner, 70–80%; Virginia, 60–65%; Spanish, 75–80%
- Other considerations
  1. If leaf spot or other diseases are a problem in the field, do not delay harvest.
  2. If weather is forecast that would delay harvest, this must be taken into account.
  3. Harvest must be done when sufficient labor and adequate equipment are available.

will maintain pods 1–2 weeks longer than plants with severe leaf spot. The shellout method has been widely used because it can be conducted in the field without further handling of pods, requires no equipment, and provides an immediate answer. When used properly in a dig, wait-to-dig manner, this method provides acceptable accuracy. With experience, a grower can use the shellout method as a simple predictor of the time at which the percentages of mature pods should reach acceptable levels for harvest (Box 5.1).

Hull-Scrape Method. The hull-scrape method for determination of the optimum time to dig peanuts was developed during the late 1980s and early 1990s and is currently accepted as the most accurate means of assessing the maturity of runner-type peanuts (Box 5.2). The method is based on the fact that the hull mesocarp (the area just beneath the tan-colored exterior of the peanut pod) changes from white to yellow to orange to brown to black as the peanut matures. As with the shellout method, the hull-scrape method begins with a representative sample from the field to be tested and consists of all pods from several plants to make a sample of approximately 200 pods. As with all testing, it is extremely important that a representative sample be obtained. The first sample should be obtained approximately 110 days after planting, and a second sample should be taken about 10 days before the predicted harvest date. Disagreement between these two samples indicates that an additional sample is needed immediately.

Execution of the method requires a color-coded chart (Plate 22) and a pocket knife or scraper to remove the tan-colored pod surface layer (exocarp). The color chart for the hull-scrape method can often be obtained from county extension offices in peanut-growing areas. Modified wet sandblasting equipment is commonly used in several states at the extension office level or the consulting laboratory to rapidly remove all the exocarp from peanuts in a sample. The point of color examination is the area surrounding the attachment point of the basal seed, commonly called the saddle area (the indention between the two seeds in the pod when the apical beak is held downward), extending toward the peg attachment point. The fresh color of the moistened pod is then matched with the color on the chart until all pods are placed. Use of the color chart gives a profile of pod development. Days until digging time are shown for the columns on the chart, and the first column from the right that contains three pods indicates the appropriate prediction. For example, if there were three pods in the column above 8–10, as in Plate 22 on the right side of the chart, the predicted time to harvest would be 8–10 days.

Severe soil moisture stress may result in substantially less or variable darkening of the middle hull area. Cool weather during late fall retards the rate of maturation. When night temperatures drop to 50°F (10°C) or below, maturation virtually comes to a stop, and color changes may never proceed to the point of a ready-to-dig situation. The decision to dig such fields must be made on the basis of the condition of the vines, peg attachment strength, and weather, particularly the potential for frost.

The prevailing attitude among peanut-production specialists is that the hull-scrape method is not accurate for Virginia- and Spanish-type peanuts and may predict a harvest time that is too early. Developers of the hull-scrape method and this author believe that with proper testing and perhaps chart modification, this method would be as acceptable for Virginia and Spanish peanuts as it is for the runner type. In some production areas, drought and the associated probability of aflatoxin development or the threat of freezing temperatures is the major consideration at harvest time, and use of any prediction method adds little to the decision-making process.

Digging

There are factors other than harvest time that affect the yield and quality of peanuts. At planting time, deep turning of the soil to reduce compaction aids in efficient digging of mature peanuts. Efficient weed control (Chapter 7) in the field and in border areas results in minimum interference during the digging and combining operations.

Digging operations are normally carried out with a digger-shaker-inverter (Plate 23). All three functions of the equipment must operate properly to accomplish the task of removing the plants cleanly from the soil and depositing them in inverted windrows (Plate 24). Equipment adjustment requires operator skill. Pitch and depth of cutting blades are critical in assuring the flow of peanut plants onto the shaker assembly. Dull or excessively pitched (upright) blades may result in soil and pods being moved forward, dislodging pods in the soil. The shaker portion of the equipment should provide sufficient agitation to remove soil as plants are lifted from the rows. Proper inverting, wherein the pods are turned up and do not contact the soil, results in the best windrow curing.

Even though pods in the air may reach temperatures high enough to cause limited flavor problems, research has shown that pods in contact with the soil reach even higher temperatures and possibly dry at a slower rate. Pods should remain dry in windrows until the average seed moisture is 18–24%. Peanuts are normally between 35 and 50% moisture at digging, and depending on weather conditions, they will dry to 18–24% in 2–4 days. If there is doubt about the percent moisture, meters are available at buying points to test pod samples. Sufficiently dried vines should ensure proper picking of pods with a combine.

Box 5.2

Hull-Scrape Maturity Testing Method

- Approximately 110 days after planting, collect three to five adjacent plants from three representative locations in a field or area of a field that can be dug in 1 day.

- Remove *all* pods from the plants from each area to obtain 180–220 pods. Repeat for each sample.

- Determine the color of the middle hull by scraping or sand blasting away the outer hull.

- Examine the middle hull color at the attachment point of the basal seed (indented or saddle area of the pod when the beak is turned downward).

- Wet pod blasters are available for grower use at some cooperative extension service offices and buying points.

- Place the pods on the peanut profile board, which is based on middle hull color (Plate 22). Keep pods wet by misting with water because color is easier to determine when pods are wet.

- Determine when to dig on the basis of slope line and harvest projection line. Find the point at which the leading edge (slope) of the sample profile crosses the projection line. Read the days until digging directly below the first column from the right that contains three pods. This date is the estimate of a 3- to 4-day range of possible harvest dates.

- A second sample should be taken about 10 days before the predicted date to verify maturity progression.

Windrow drying below 18% will usually increase harvesting losses, which will exceed the cost of fuel and electricity for mechanical drying. Complete drying in windrows should generally be avoided. Rainfall during windrow drying of pods with about 20% moisture may promote mold growth and an overall reduction in milling quality. Extended rainfall will cause substantial field losses resulting from deterioration of pod-to-peg attachments.

Combining

When plant and peanut moisture have been adequately reduced, pods are harvested directly from the windrows with a combine (Plate 25). Whole plants are interlaced in windrows, providing a continuous ribbon of vines that is pulled into the combine where pods are removed, cleaned, stemmed, and conveyed into a bulk bin on the combine. It is essential that the combine pickup reel speed be synchronized with the combine ground speed so that the ribbon of plants flows continuously into the machine (Plate 26). Improper adjustment will result in the windrow being torn apart and pods stripped from the vines before entering the machine, or the combine may overrun and drag the windrow. The picking fingers on the pickup head should be adjusted so that they pick up the windrow without digging into the soil.

The adjustment of picking cylinder speed and clearance inside the combine can have a major effect on the picking efficiency and the number of LSK, which are lower in value (grade) than in-shell peanuts. LSK are more susceptible to damage and deterioration in storage and generally have a higher potential for aflatoxin contamination than other kernel categories. Adjustments to reduce the percentage of LSK should be based on combine manufacturer recommendations and may need to be performed during each day of operation since plant conditions change from morning to night.

The first separation of peanuts from all other material occurs during combining (Box. 5.3). Material that may be picked up with the peanuts includes soil, stems, grass seed, nutsedge tubers, glass, peanut shells, and many other items. Foreign material impacts several areas that affect quality, beginning with airflow restrictions and uneven moisture distribution during curing. Foreign material at 5% and above results in a deduction in the value of farmer's stock peanuts brought to market. During storage, foreign material interferes with airflow, reducing the ventilation that is necessary for the removal of moisture from the warehouse. The presence of foreign material in shelled stock and finished products is obviously not desirable and may result in consumer complaints and litigation. Separation of these foreign materials is greatly affected by the airflow in the combine. Initial combine adjustments should always be made according to manufacturer's recommendations. Field observations should be made of foreign material remaining in the harvested pods and of the number of pods being blown out the rear of the combine. These observations may indicate a need for further adjustment of the combine.

During picking, pods are separated from most of the vines and other foreign material and are generally conveyed into a bin on the combine (Plate 27). Airflow is usually the method of conveyance, and the proper flow of air is critical for minimizing physical damage. Significant damage (50% in some studies) may occur during combining, depending on the moisture content of the crop and combine cylinder speed. This physical damage may expose seed to further damage from dust, dirt, and insects and may also result in damaged or ruptured cells, leading to rapid reduction in flavor quality. Combine bins are constructed for direct dumping into peanut drying wagons of approximately five-ton capacity (Fig. 5.1). Dumping should be done so that peanuts are of uniform depth when the wagon is full.

Curing

"Curing" is the process of water removal such that peanut biochemistry and physiology are optimum for food quality. Proper curing is essential for safe storage, milling quality, and development of biochemical factors required for optimum roast flavor quality in peanuts. Curing and its interaction with maturity make up the single most important factor in establishing the basic flavor quality of peanuts after harvest. This critical step in obtaining peanut quality is often overlooked because of the rush to get the crop harvested. Curing is most often accomplished in drying wagons (Fig. 5.1) by blowing heated air through the perforated bottom of each wagon. Harvesting is intense, and wagon availability is sometimes the limiting factor in keeping the process moving. Thus, curing may often be attempted with excessive heat to decrease drying time, which results in an unacceptable peanut product. "Drying" may be considered water removal with little concern for the finer points of biochemistry that are important for optimum food quality. Because of common usage, the terms curing and drying are often used interchangeably.

The potential for loss of peanut quality during curing is related to original quality, moisture content control, and damage before curing. Original quality is determined by growing conditions, pod maturity, windrow conditions, and the care taken during the harvesting operation. Moisture control is determined by the temperature and humidity of air used to dry the pods. When the curing process begins, a drying zone is initially created in the pods at the bottom of the wagon, and this zone moves upward as the curing process continues. The drying process must be continued long enough for the drying zone to move entirely through the peanuts on the wagon. If air temperature is too high and/or relative humidity is too low, the bottom layers of peanuts will be overdried, and losses in quality and marketable weight will result. If air temperature is too low, relative humidity too high, and/or airflow rate too low, the drying rate will be too slow, and a different set of quality losses will

---

Box 5.3

Reducing Foreign Material, Loose-Shelled Kernels, and Pod Damage During Combining

- Combine windrow-cured peanuts at 25% or less moisture as determined by an electronic moisture meter.

- Adjust pickup reel speed to correspond to ground speed so that plants flow continuously into the combine without tearing windrows apart.

- Picking cylinder speed and clearances should be adjusted daily to manufacturer specifications and vine conditions during harvesting. The need for adjustment may be indicated by a high number of loose-shelled kernels in the bin.

- Large amounts of foreign material going into the bin with the pods and a high number of pods being blown out the back of the combine may indicate a need for combine adjustment.

occur. Drying zone thickness is a function of airflow rate. High airflow rates result in thick drying zones with more uniform drying parameters. However, these high rates are likely to force unsaturated air through and out of the peanut load, thus reducing the energy efficiency of the process. Low airflow rates produce narrow drying zones with a wide range of moisture contents from the top to the bottom of the load.

Drying recommendations vary among the peanut-growing regions. However, all current airflow rate recommendations are a compromise between high energy efficiency versus high drying rates and uniformity of drying. Some consolidation of these practices is assumed in the following discussion of recommended practices. Box 5.4 provides a summary of general guidelines.

Moisture of peanuts should be 25% or less, and peanuts should not exceed 5 feet in depth in the drying wagon. Peanuts should be leveled because mounding will cause nonuniform depths and peanuts at the peak of a mound will not dry properly. Ambient or heated air should be applied within a few hours of combining to prevent hot spots and growth of bacteria and fungi. The maximum thermostat settings of the dryer and the allowable rise in temperature are influenced by the actual relative humidity and ambient air temperature. Because of regional differences, it is essential that extension service guidelines for each growing area be followed. Generally, airflow should be no less than 50 cubic feet per minute for each square foot of drying floor. The thermostat should not be set to exceed 95°F (35°C), and temperature rise above ambient, which is controlled by burner orifice size and gas pressure, should not be more that 15°F. Temperatures in the wagon plenum (the area below the perforated floor) should be checked hourly, and consideration must be given to temperature rise relative to maximum and minimum daily temperature. Temperature in the wagon can be measured by drilling a small hole in the plenum opposite the heated air entry and inserting a stem thermometer. Infrared thermometer "guns" may be used to check the external temperature of the drying wagon as an indicator of heated air temperature.

"Curing" is the process of water removal such that peanut biochemistry and physiology are optimum for food quality.

Because peanuts are dried to a predetermined average moisture content, it is important that moisture samples be taken with a probe from at least three locations in the wagon or bin dryer. Each sample should contain peanuts from all depths in the wagon. A combined sample is obtained by riffle dividing the composite materials from several probes. The initial moisture of each load should be checked, and sampling frequency should be adequate to prevent overdrying of peanuts. At low moisture (13% or less), samples should be taken at least hourly and more frequently if drying is progressing rapidly. Under average conditions, mechanical curing should be terminated when peanuts on the wagon reach 11.0–11.5% moisture. Curing will continue because of the "coasting effect," which lowers the moisture content another 1–1.5% below the cut-off point. This effect is caused by the curing temperature and the unequal moisture content of the shells and kernels. A final average kernel moisture

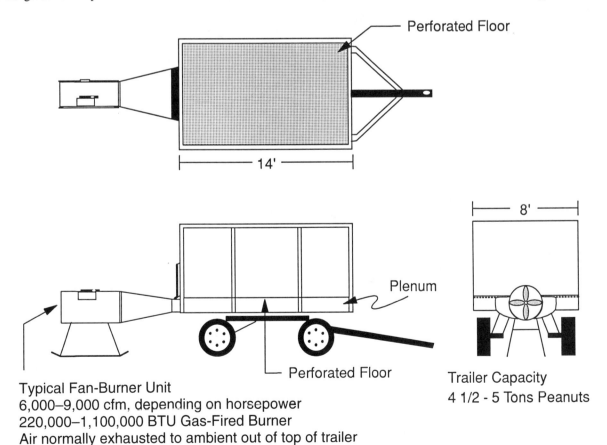

Perforated Floor

14'

Plenum

Perforated Floor

Typical Fan-Burner Unit
6,000–9,000 cfm, depending on horsepower
220,000–1,100,000 BTU Gas-Fired Burner
Air normally exhausted to ambient out of top of trailer

8'

Trailer Capacity
4 1/2 - 5 Tons Peanuts

Fig. 5.1. Peanuts are dried in drying wagons with heated forced air. cfm = cubic feet per minute.

content of 9–10% is desired for marketing and safe storage.

Drying air temperature that is too high and overdrying are two of the most common deviations from recommended drying practices. These will increase the percentage of split kernels when peanuts are shelled. Increases of 5% or more are common, depending on the interactions of drying air temperature, maturity, relative humidity, and final moisture content. If peanuts are overdried, the percentage of split kernels increases, the marketable weight of the load is reduced, and the value decreases. Often, overdrying or drying too quickly with high temperatures weakens the peanut hulls, which may result in more peanuts being shelled when samples are taken for grading. Grading may then indicate an incorrectly high LSK percentage. High LSK values decrease the value of the load, and LSK are much more likely to undergo quality deterioration, *A. flavus* invasion and aflatoxin contamination, or insect invasion during later storage (Chapters 8 and 13).

During the late 1940s and early 1950s, it was recognized that curing affected peanut flavor and that excessively high temperatures resulted in off-flavors. Immature peanuts are more susceptible to off-flavors, and specific volatile compounds are produced in peanuts, indicating that they have been improperly cured. A recently developed organic volatile meter has been suggested as an addition to the present grading system. It has the potential to identify loads of peanuts with freeze damage or off-flavor caused by high-temperature curing. The descriptive flavor term associated with damage of either type is "fruity fermented," and the flavor is most often detectable in the smaller peanuts in a load. At present, off-flavor problems are more subtle than LSK, split kernels, and other physical damage since they are often not obvious until manufacturing.

During the curing process, peanuts are dried to an average moisture content of about 10%, which means that some kernels are drier than 10% and others contain more moisture. The moisture content range of kernels in a lot is related to initial moisture and maturity of individual kernels. Problems of overdrying have been discussed; however, peanuts that are too moist present a completely different problem as they move through storage,

---

Box 5.4

General Guidelines for Curing Peanuts

- The moisture content of peanuts from windrows should be 25% or less.

- Peanuts should be level across the drying wagon to provide a uniform depth no greater than 5 feet.

- Airflow should be applied to loaded drying wagons as soon as possible.

- Dryer thermostats should be set no higher than 95°F (35°C).

- Airflow should be no less than 50 cubic feet per minute per square foot of drying floor.

- Moisture sampling should be across all depths and more frequent as moisture approaches 13% or less.

- Drying should be terminated at 11.5–11.0% because the "coasting effect" will lower percentages another 1–1.5% to provide a moisture average of approximately 10%.

"Drying" may be considered water removal with little concern for the finer points of biochemistry that are important for optimum food quality.

shelling, and marketing. Immature peanuts may contain twice as much moisture as mature peanuts at harvest. Although the difference decreases during curing, it remains even through the storage period. After normal drying and 5 months' storage, some immature peanuts have been found with a moisture content as high as 17%. This is a level unacceptable for any length of storage. Harvest practices, crop maturity, curing, and variable moisture content are all related to quality maintenance in farmer's stock (in-shell) storage.

Storage

Cleaning

Harvested peanuts contain a wide variety of foreign materials that may significantly affect storability of farmer's stock peanuts. Elimination of as much foreign material as possible will help maintain peanut quality during warehouse storage. Included in the list of materials are soil, rocks, sticks, immature and shriveled pods, peanut plant parts, weed fruit, nutsedge, corn cobs, metal, and even such miscellaneous items as glass and firearm shell casings. The types and quantities of material depend upon adjustment and operation of harvesting equipment, cultural practices, soil type, field history, and other obvious factors. Although not considered foreign material, LSK are often dirty, moldy, mechanically damaged, or insect damaged, and they deteriorate more rapidly during storage than in-shell peanuts. When insecticides are applied to peanuts entering storage, they are deposited directly onto the LSK without the protection of shells. Small, shriveled pods generally have high moisture contents and often mold during storage. As peanuts are loaded into warehouses on conveyors, foreign materials such as soil, sand, and fine material tend to concentrate at the discharge point. This concentration of material restricts airflow and prevents proper moisture movement as peanuts continue to cure during storage.

Most peanuts are not cleaned before storage because of the initial investment in equipment and the potential for delay in the rapid harvesting process. The Federal State Inspection Service requires cleaning of farmer's stock loads with greater than 10% foreign material. Cleaning equipment and methods to remove some of the foreign material have been developed and include sand screens, belt screens, and other cleaners capable of rapidly removing the wide range of material found in peanuts. Recent peanut industry (National Peanut Council) recommendations state that prior to the admission of farmer's stock peanuts into warehouses or the shelling system, loads with excessive foreign material (more than 4%) or excessive LSK (more than 5% suggested) should be passed through a cleaning system. Special attention should be given to the removal of LSK, high-moisture material, and dirt, since these materials will likely increase the risk of insect damage and mold, potentially leading to aflatoxin contamination, during storage. LSK from farmer's stock peanuts may meet edible quality requirements at this point; however, after a period of storage, they may not.

Farmer's Stock Storage

Peanuts are a semiperishable crop and thus are subject to quality deterioration during storage. It is often stated that peanut quality cannot be improved during storage; it can only be maintained. Maintenance of high quality can occur only when incoming peanuts have high quality and proper moisture and are clean. Peanuts lose moisture during the storage interval, which may last from September to April or May, depending on when they are delivered to shelling plants. Moisture control is the primary factor in storage of farmer's stock peanuts.

Farmer's stock peanuts are variously stored in tanks, bins, concrete silos, and flat-type warehouses with natural or mechanical ventilation systems. Most peanuts in the United States are stored in flat, uninsulated steel warehouses usually about 85 feet wide and 12–24 feet high (at the eaves) with a roof slope of at least 33° (Plates 28 and 29). Natural ventilation is usually supplied by ridge and eaves vents made as large as possible to provide maximum ventilation. Vents should always be kept free of obstructions, and peanuts must not be loaded against the vents. Low cost, low maintenance, and low energy utilization are reasons for using natural ventilation systems. Mechanical ventilation consists of propeller-type fans that have the capacity to draw at least one air change through the warehouse overspace (the area above the peanuts) every 3 minutes. Large inlet louvers are installed opposite the fan end to allow sufficient air to enter the warehouse. Mechanical ventilation usually eliminates dead air in the overspace, but maintenance and inspection are needed to ensure proper operation. Condensation and subsequent growth of microorganisms are likely if the mechanical system fails at a critical time. Periods of condensation and mold growth can occur when there is a rapid decrease in ambient temperature during the first part of the storage period.

Ventilation systems are used to help maintain the quality of peanuts by removing excessive heat and moisture and equalizing moisture content and temperature in the mass of stored pods. The systems function to keep the condition of air in the peanut pile within certain limits to prevent moisture migration and condensation inside the warehouse. As mentioned previously, the range of pod moistures comprising a 10% average may be quite broad (5–15%), and temperatures at loading range from 70 to 95°F (21 to 35°C). During storage, pod moistures equilibrate and temperature of the pods decreases because of the cool air being drawn into the warehouse by the ventilation system. The cool air displaces warm, moist air in the peanut pile, and as the warm air rises, it can reach the dew point and condense if not exhausted before it comes in contact with a cool surface, such as the warehouse roof. Condensation may result in microbial growth over the surface of the entire pile of peanuts, or condensation drip may cause large, molded clumps of peanuts, commonly called soldiers (Plate 30). In the early part of the storage season when moisture and temperature are high, ventilation fans are operated continuously, except during fumigation, insecticide application, or long periods of very humid conditions (greater than 90% relative humidity). As a point of reference, as peanuts in a 5,000-ton warehouse cure from 10 to 7% moisture during the storage season, approximately 36,000 gallons of water must be removed by the ventilation system. Most water removal occurs during the early part of storage as the temperature of the peanuts decreases.

During the middle of the storage season, ambient temperatures are usually low and relative humidity averages are nearly ideal, except in some very arid areas. Serious milling losses may occur if seed are dried to less than 7% moisture or temperature at the time of shelling is less than about 45°F (7°C). Primary concerns during this time are the removal of radiated solar heat and the maintenance of proper temperature and moisture content of the peanuts.

During late-season storage, condensation may again become a problem as warm, moist outside air comes into contact with the cool peanuts. Ventilation systems are utilized to maintain the peanut temperature slightly below the average daily temperature.

> # Peanut quality cannot be improved during storage; it can only be maintained.

As the temperature increases and insects become active, infestations can occur quickly. Supervision and proper treatment are needed to control insect problems. To preclude insect problems, peanuts are generally sprayed with malathion as they enter storage.

Tips for storing peanuts are summarized in Box 5.5.

Shelled Stock Storage

Peanuts removed from storage are transported to a shelling plant where they are graded, cleaned, shelled, and separated into commercial grade sizes. Shelling operations remove the hull (a natural seed-protection barrier) and may impart physical damage to the seed. Shelled peanuts therefore tend to deteriorate faster than unshelled peanuts and usually need to be stored under controlled conditions. Generally, shelled peanuts can be held for a year or more at 33–41°F (1–5°C) and 55–70% relative humidity without serious quality loss. The moisture content of the peanuts should be no more than approximately 7%, and slightly lower values provide additional storage potential. The goals of refrigerated storage of shelled peanuts are to maintain relative humidity in the proper range and supply an odor- and chemical-free environment. Shelled peanuts absorb chemicals such as ammonia, which is used in some cooling systems, and will absorb odors from foods such as onions, garlic, cantaloupes, herbs, meat, cheese, and chocolate. Storage of peanuts with these chemicals and products should be avoided.

Box 5.5

Tips for Storing Peanuts

- Peanuts should be cleaned before storage to remove soil, loose-shelled kernels (when content is greater than 5%), and high-moisture foreign material (when content is greater than 4%).

- Peanuts should be treated with a recommended insecticide prior to storage.

- Peanuts lose moisture during farmer's stock storage, and this moisture must be eliminated from the warehouse if quality is to be maintained during storage.

- Excessive heat buildup must be prevented during storage.

- In mechanically ventilated warehouses, fans should be run continuously during the early part of storage.

- Warehouses should be examined frequently for moisture and insect problems, which can quickly reduce quality.

Measurement of Quality

Grading

As stated earlier, quality is, in part, a measure of consumer acceptability. It must be realized that there are many different consumers in the peanut system. Peanut producers are even considered consumers when they buy peanut seed for planting. One of the quality measures used for seed is germination percentage. Quality may be measured in different ways for different consumers.

One of the most important methods of measuring quality is grading as administered by the Fresh Products Branch of the U.S. Department of Agriculture's (USDA) Agricultural Marketing Service. Through cooperative agreement with peanut-producing states and the USDA, states employ peanut inspection personnel to make standardized grade determinations of peanuts. Domestic peanuts are graded at least twice by the Inspection Service. The first inspection is on farmer's stock peanuts delivered by producers to shellers (handlers). This sample is removed from a load by using a pneumatic probe (Plate 31). The second inspection for grade is on milled peanuts after they have been shelled, sized, and cleaned for shipment. The Inspection Service does not specify what grade and quality factors will be determined, but it does specify how these factors will be determined.

> **High farmer's stock grades are critically related to proper harvest date, harvesting methods, and curing.**

High farmer's stock grades are critically related to proper harvest date, harvesting methods, and curing. In grading, quantitative measurements are made that are affected by most of the factors discussed in this chapter. In farmer's stock grading, the percentage of SMK (seed that ride a 16/64-inch slotted screen) is determined. This percentage is the factor with the greatest influence on price and is directly related to proper harvest time. Percentage of split kernels and moisture content, which are related to curing conditions, are also evaluated. Foreign material, LSK, and freeze damage, also determined in grading, are related to maturity, windrow conditions, and combining operations. Also, as dictated by the USDA peanut-marketing agreement, all of the seed from each sample must be examined for visible growth of the aflatoxin-producing fungus *A. flavus* (Chapter 13). Seed lots that are found to contain visible growth of *A. flavus* are classified as Segregation III. Seed lots with no visible growth of *A. flavus* but with more than 2% damaged seed or more than 1% concealed damage (i.e., damage not visible unless the seed is split) caused by mold or decay are classified as Segregation II. All others are classified as Segregation I. Segregation III peanuts are used for oil, and the meal is restricted to nonfood uses.

Peanut quality is the result of steps taken at all stages from the planting of the peanut seed to the time the finished product is consumed. At each stage, improper conditions or actions may eventually lead to an inferior product. Often these missteps are additive; that is, minor deviations from acceptable practices in more than one area accumulate to produce an unacceptable product. It is imperative that proper cultural practices and handling procedures be used throughout the entire process.

Flavor Quality

Some inherent characteristics of peanuts, such as maturity, affect flavor, and some external factors, such as high-temperature curing and improper storage, are the direct causes of obvious flavor problems. Flavor quality is affected by the interactions of inherent characteristics and external factors. The varying effects of environment on the precursors of flavor development in different market types and cultivars should produce the potential for a wide range of normal flavor characteristic combinations; however, this range is relatively narrow. The influence of these interactions between environmental and cultural conditions and cultivars is exhibited in the intensity ratings of various flavor descriptors. The range of off-flavors inherent in peanuts is also relatively narrow but must be extended greatly to accommodate the numerous volatile products to which peanuts may be exposed.

Flavor has been defined by a number of researchers and manufacturers in lists of terms that commonly include flavor descriptors such as roasted peanutty, sweet aromatic, woody, raw beany, painty, and fruity fermented plus the four tastes sweet, sour, salty, and bitter (Box 5.6). Lists of terms like these have been used as the basis for descriptive analysis of peanuts in studies, which have included research on the effects of maturity and curing of peanuts, the effectiveness of the volatile organic meter, harvest dates, peanut butter formulations, and product packaging materials.

The various lists of descriptors have been used to train numerous people throughout the peanut industry as a means of communication to help solve problems that occasionally arise during the production, marketing, and manufacturing operations. Further, the lists have been used to relate specific growing, curing, handling, storage, and manufacturing problems to specific identifiable terms. These associations are not always absolute, and care must be taken that assumptions are verified by chemical analysis when possible.

Quality—A Continuous Process

Because of the wide diversity of occupations involved in the total production of peanut products, there is often little regard for overall peanut health management. Peanut quality, measured

---

Box 5.6

Peanut Flavor Descriptors

Aromatics
Roasted peanutty
Raw beany
Dark roast peanut
Woody/Hulls/Skins
Cardboard
Painty
Other (burnt, green, earthy, fishy, grainy, chemical/plastic, sweet fruity/fermented)

Tastes
Sweet
Sour
Bitter
Salty

Feeling Factors
Astringent
Metallic

by many and various standards, is a continuous process. Seedsmen, farmers, buying-point operators, warehousemen, handlers, and manufacturers are all important to the delivery of high-quality peanuts and peanut products to their various consumers. Extension agents, peanut specialists, consultants, scientists, and other advisers also play important roles in the continuous process. It is possible that one error in judgment or operation will not result in a noticeable decrease in peanut quality, but the potential increases with each error. Management activities related to seeding rate, leaf spot spray cycle, irrigation timing, and harvest date will affect peanut health and ultimately peanut quality. Management activities that determine curing temperature and storage conditions are crucial for maintaining quality. All other routine operations that result in handling the peanuts have an impact on maintaining their quality.

> ## Quality is a continuous process. Every step is critical to obtaining or maintaining quality.

Unique roasted flavor and nutrition are the main factors in final sales of peanut products, but the real driving force for the peanut industry is economics. As technology and understanding of factors affecting peanut quality advance, there should be less and less compromise between peanut quality and economics. Ultimately, peanuts and other agricultural products are sold on the basis of quality and wholesomeness, and loss of sales in other commodities has occurred when quality or wholesomeness was questioned. Past and current research indicates that production and processing practices that result in high yields and financial gain usually have strong relationships to maximum quality potential. It is essential that all parts of the peanut industry learn and apply good management practices.

Quality is a continuous process. Every step is critical to obtaining or maintaining quality.

Selected References

Anonymous. 1989. Peanut Industry Good Management Practices. National Peanut Council, Alexandria, VA.

Baldwin, J. A., Beasley, J. P., Colburn, A. E., Glover, J. W., Hartzog, D. L., Kvien, C., Sholar, J. R., Sullivan, G. A., Swann, C. W., Whitty, E B., Williams, E. J., and Wright, F. S. 1990. Peanuts: A Grower's Guide to Quality. Planters LifeSavers Company, Winston-Salem, NC.

Davidson, J. I., Jr., Whitaker, T. B., and Dickens, J. W. 1982. Grading, cleaning, storage, shelling, and marketing of peanuts in the United States. Pages 571-623 in: Peanut Science and Technology. H. E. Pattee and C. T. Young, eds. American Peanut Research and Education Society, Yoakum, TX.

Dickens, J. W., and Johnson, L. W. 1987. Peanut grading and quality evaluation. Pages 36-47 in: Peanut Quality—Its Assurance and Maintenance from the Farm to End-Product. E. M. Ahmed and H. E. Pattee, eds. Fla. Agric. Exp. Stn. Bull. 874.

Fletcher, M. M. 1987. Evaluation of peanut flavor quality. Pages 60-72 in: Peanut Quality—Its Assurance and Maintenance from the Farm to End-Product. E. M. Ahmed and H. E. Pattee, eds. Fla. Agric. Exp. Stn. Bull. 874.

Johnson, W. C., III, ed. 1987. Georgia Peanut Production Guide. Cooperative Extension Service, University of Georgia College of Agriculture, Athens.

Sanders, T. H., Blankenship, P. D., and Smith, J. S., Jr. 1987. Peanut quality in curing and storage. Pages 29-35 in: Peanut Quality—Its Assurance and Maintenance from the Farm to End-Product. E. M. Ahmed and Harold E. Pattee, eds. Fla. Agric. Exp. Stn. Bull. 874.

Sanders, T. H., Schubert, A. M., and Pattee, H. E. 1982. Postharvest physiology and methodologies for estimating maturity. Pages 624-654 in: Peanut Science and Technology. H. E. Pattee and C. T. Young, eds. American Peanut Research and Education Society, Yoakum, TX.

Young, J. H., Person, N. K., Jr., Donald, J. O., and Mayfield, W. D. 1982. Harvesting, curing and energy utilization. Pages 458-485 in: Peanut Science and Technology. H. E. Pattee and C. T. Young, eds. American Peanut Research and Education Society, Yoakum, TX.

Craig Kvien
Department of Agronomy
University of Georgia, Tifton

CHAPTER SIX

Physiological and Environmental Disorders of Peanut

In peanut, as in other crops, most physiological disorders are caused by stresses that are either biotic or abiotic. Biotic stresses are caused by insects, bacteria, fungi, viruses, nematodes, and weeds. Stand reduction caused by fungal and insect pathogens is one symptom of biotic stress. Another is the reduction in photosynthesis from the loss of green leaf area caused by leaf spot, rust, or foliage-feeding insects. Weeds can cause biotic stresses by competing with peanut plants for water, nutrients, and light. Sugars and amino acids may be removed from the plant by nematodes or sucking insects. Turgor reduction may occur as a result of root damage caused by nematodes and other soilborne pathogens or as a result of tissue consumption by root-, pod-, and foliage-feeding pests. Most of the effects of biotic stresses are covered in detail in other chapters.

This chapter focuses on abiotic stresses related to temperature, water, radiation, chemicals (pesticides, nutrients, salts, and gases such as ozone), and other physical conditions such as wind and equipment traffic. Stresses may have direct or indirect effects on plants. An example of a direct effect is the sudden exposure of a plant to subfreezing temperatures. Ice crystals form inside the cell, tearing the plasma membrane and killing the cell. An indirect effect of chilling stress could occur if a plant is exposed to temperatures just above freezing. If the stress is brief, the plant may fully recover. However, if the chilling stress persists for many hours, the plant's membrane lipids (fats) may solidify. This solidification leads to varying degrees of inactivation among certain membrane-bound enzymes, resulting in the accumulation of toxic intermediate compounds that may damage the plant and cause an indirect stress injury. High temperatures can also produce direct injury, but this condition is associated with a water deficit that causes a secondary stress injury to the plant.

Plants adapt to stresses in two basic ways: avoidance and tolerance. In general, avoidance mechanisms insulate plant cells, allowing them to function normally by excluding the stress with physical, chemical, or metabolic barriers. For example, one avoidance adaptation to drought stress is the thickening of a plant's cuticle (the waxy layers on the top and bottom of the leaf). Thicker cuticles are less permeable to water and help the plant maintain turgor and avoid excessive water loss. Tolerance is the other way plants adapt to stress.

Plants tolerate stress by coming to thermodynamic equilibrium with the stress. For example, a drought-tolerant plant survives by maintaining its cells at low water potential until the stress is relieved. During high temperatures, a heat-tolerant plant is able to function at a high plant temperature, whereas a plant avoiding heat stress would have mechanisms to keep its temperature lower than ambient.

In this chapter, brief descriptions of how different abiotic disorders of peanut affect the plant and suggestions for minimizing their effects are given.

> ## Plants adapt to stress in two basic ways: avoidance and tolerance.

Disorders induced by temperature, water, radiation, pesticides, nutrients, gases, and equipment are common in peanut as well as in other crops. These disorders interact with each other and with various biotic disorders, sometimes creating diagnostic puzzles. Like biotic pathogens, the causes of some abiotic stresses (water surplus or deficiency, nutrient supply, equipment traffic, and pesticide injury) are more easily managed than others (radiation, temperature, and ozone damage). Proper selection of peanut cultivars with resistance or tolerance to the abiotic stresses likely to be encountered is one of the best ways to manage these disorders. Performance of cultivars that are well adapted to the normal stresses imposed in a given growing area is often checked in tests performed by the research or extension services of the various states. Lists of adapted cultivars and their performance are generally available in the county extension offices.

Temperature Stresses

High and low temperatures can impair peanut growth and development (Chapter 3). Because peanut is of subtropical origin, it is sensitive to low temperatures. However, peanut yields can also be adversely affected by very high temperatures. The influence of temperature on peanut physiology is complex and often confounded with other stresses that are associated with temperature stress. For example, high temperatures are often associated with drought and high light intensities. As soil water becomes limited, the tiny pores (stomata) in the leaf close, decreasing the cooling effect of water evaporation from the leaves (evapotranspiration). This causes the temperature of the leaves to rise above the ambient temperature. Peanut can survive at temperatures greater than 95°F (35°C) and has been known to withstand temperatures as high as 120°F (49°C) for short periods. However, photosynthesis is reduced at these high temperatures, and if night temperatures are above 95°F (35°C), flower set and peg formation are inhibited. The best temperature for peanut photosynthesis and dry matter production is approximately 86°F (30°C).

Research with Florunner has shown that soil temperatures of about 80°F (27°C) generally produce larger seed that take longer to mature than seed formed at higher soil temperatures. Soil temperatures above 95°F (35°C) and below 70°F (21°C) may result not only in production of smaller pods but also in fewer pods than would be expected at optimal temperatures. Figure 6.1 illustrates the relationship between temperature and peanut seed size distribution. Note that fewer large seeds (jumbo) were produced when soils were heated and higher percentages of medium and small seed were evident. Figure 6.2 shows the relationship between temperature and maturity. At lower temperatures, maturity is delayed, and the percentage of mature pods (classes 6 and 7) is reduced. The effect of maturity on seed size distribution is illustrated in Plate 32. The more mature peanuts have a greater percentage of large kernels than those that are less mature.

Low-Temperature Stress

The effect of low temperature on the crop will depend on when the stress occurs during the growing season. Early season low-temperature stresses delay germination, decrease seedling vigor, and increase the likelihood of plant loss from seedling diseases.

In addition to temperature, the availability of water and oxygen are the primary external factors affecting germination. Soils that are cooler and wetter than ideal will subject seedlings to chilling stress, which can result in a higher percentage of abnormal plants, germination failure, slowed growth, and greater susceptibility to disease. Seeds germinated under cool conditions are known to produce a higher percentage of plants with short, stubby, curled roots and thick hypocotyls. Low-temperature stress (below 65°F [18°C]) is most harmful to the crop when it is first planted and the seeds are absorbing (imbibing) water. After the seeds have imbibed water and are swollen to full size (within 8–24 hours), they are less sensitive to low-temperature stresses.

> **The best temperature for peanut photosynthesis and dry matter production is approximately 86°F (30°C).**

Optimal germination temperature is cultivar dependent because there are genetic differences among cultivars in their ability to tolerate cold. In general, soil temperatures of 65–90°F (18–32°C) are adequate for good germination (Chapter 3). Outside either temperature extreme, physiological and pathological effects will have an adverse impact on germination success.

Later in the growing season, low temperatures will reduce the photosynthetic rate and dry matter production of the plant. Little growth or photosynthesis occurs at temperatures below 60°F (15°C). Peanut plants are very sensitive to cold and may be killed by temperatures at or slightly below freezing. As the temperature approaches freezing, the cell membranes lose their integrity, resulting in the malfunction of enzyme systems. Temperatures below freezing cause formation of ice crystals, which tear cell membranes and kill the cells. However, unless

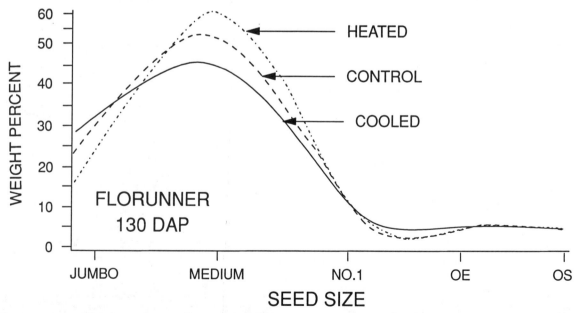

Fig. 6.1. Effects of soil temperature on seed size distribution. The three temperatures are the average soil temperatures from 28 days after planting until harvest. Heated = peanuts grown at 84°F (29°C); control = peanuts grown at 76°F (24°C); and cooled = peanuts grown at 71°F (21°C). (Courtesy T. H. Sanders)

soil temperatures fall near freezing, the pods on frost-damaged plants are usually undamaged, although pod stem strength will decline more quickly on damaged plants. Therefore, harvest of the damaged plants should begin when temperatures are expected to rise above freezing for a period long enough to permit safe curing in windrows. Chilling damage to high-moisture seed during curing in the windrow may cause flavor deterioration and a decrease in seed quality.

> # Chilling damage to high-moisture seed during curing in the windrow may cause flavor deterioration and a decrease in seed quality.

The most common late-season problem caused by low temperatures is freeze or chilling damage to high-moisture seed in the windrow. Leakage of cell solutes, caused by plant membrane damage, and anaerobic respiration (respiration without free oxygen) both result from low-temperature stress to the seed. Changes in enzyme systems lead to increased concentrations of organic volatiles such as alcohol and ethanol and can cause seed and flavor deterioration.

Box 6.1 contains tips on managing low-temperature stress.

High-Temperature Stress

Heat stress can have a variety of physiological effects on the peanut crop, including inhibition of photosynthesis, disruption of respiration, changes in membrane permeability, and interference with nutrient movement. Of these, the photosynthetic process is the most easily affected by heat.

Heat tolerance differs among cultivars and varies with the environment in which plants are grown. For example, subjecting a plant to a gradual water stress can improve the plant's ability to tolerate heat. This conditioning effect will disappear after the stress is relieved, indicating that the "hardening" process is reversible.

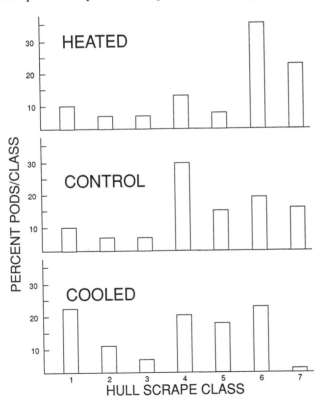

Fig. 6.2. Effects of temperature on maturity distribution of peanuts. The three temperatures are the average soil temperatures from 28 days after planting until harvest. Heated = peanuts grown at 84°F (29°C); control = peanuts grown at 76°F (24°C); and cooled = peanuts grown at 71°F (21°C). Maturity (hull scrape) classes: 1 = least mature to 7 = most mature. (Courtesy T. H. Sanders)

Box 6.1

Tips for Reducing the Effects of Low Temperatures

• Plant only in soils that are warmer than 65°F (18°C). Planting dates should be selected to avoid early season and late-season chilling stress. When forced to plant in cool soil, use only high-quality seed.

• Cultivars differ in their ability to tolerate cold temperatures. When applicable, temperature tolerance should be used as one of the cultivar-selection criteria.

• Do not dig if freezing temperatures are expected. High-moisture peanuts in the windrow will be damaged. Do not leave high-moisture peanuts in the windrow if a frost is expected.

• If a heavy frost occurs before the crop is mature, it is unlikely that the peanut plants will recover. Harvesting as soon as possible after a freeze is advised. However, harvesting should proceed after the threat of freezing temperatures has passed so as to avoid freeze damage to the windrowed peanuts. In some cases, peanut plants can recover from a short, late-season cold spell of above-freezing temperatures that is followed by many warmer days where temperatures exceed 60°F (16°C).

Box 6.2

Tips for Reducing the Effects of High Temperatures

• Regional area temperature records should be noted and planting dates selected so as to avoid excessively high temperatures (above 95°F) (35°C) during the critical periods of germination and peak flowering.

• If available, irrigation can be used to reduce damage caused by high temperatures. Adequate soil moisture allows the peanut plant to transpire and thus lower its leaf temperature. One irrigation application of 1.5 inches is preferable to two or more applications in smaller amounts.

• Provide conditions favorable for the peanut crop to establish and maintain a good root system. Healthy roots improve water uptake and translocation to the leaves, thereby reducing the negative effects of high temperatures. Root function can be strengthened by improving soil tilth, rotating with a grass crop, decreasing soil compaction, maintaining subsoil moisture, and controlling soilborne insects.

Early in the season, unusually high soil temperatures will adversely affect seed germination. Excellent germination usually occurs at temperatures up to 90°F (32°C); a sharp decline in germination occurs at temperatures exceeding 95°F (35°C).

In addition to reducing the seed germination rate, excessively high temperatures will increase the percentage of abnormal seedlings. Seedlings with black root tips and necrotic areas indicate damage from high soil temperature. After germination, overheating of the soil surface can burn the hypocotyl and, if the intensity is sufficient, may kill the plant.

Later in the season, high-temperature stress can inhibit photosynthesis, slow dry matter accumulation, and reduce fruit set. Redistribution of nutrients can also occur as a result of high temperatures. For example, when night temperatures exceed 95°F (35°C), flower set decreases, thus limiting the number of developing pods, which in turn makes available more assimilates for leaf and stem growth.

Very high temperatures above 95°F (35°C) will increase respiration rate, decrease the efficiency of many enzyme systems, and delay development of the peanut crop. As with excessively low temperatures, high temperatures are known to cause plant cell membrane damage and electrolyte leakage. Excessively high soil surface temperatures can burn peg tips, resulting in reduced pod set.

Although no one can control the weather, there are management practices (Box 6.2) that can be employed to help reduce the risk or effects of high temperatures on the peanut crop.

Water Stresses

Water-related stresses can arise from water deficits or water excesses. In most peanut-growing regions, water deficits are a more common problem than waterlogging.

Water Deficit (Drought)

Visible signs of a drought-stressed peanut crop typically include wilting, with the leaves turned so as to display more of their silvery undersides. When the leaves are turned this way, the plant conserves water by reflecting more light. In addition, the wilted leaves are not perpendicular to the sun, and radiant energy input into the canopy is thereby reduced. Both mechanisms reduce water loss by transpiration.

Drought can produce either a harmless, reversible effect or more injurious stresses that permanently affect growth, metabolism, and membrane structure. Plant processes vary in their sensitivity to water shortages. Cell growth, cell wall development, and protein and chloroplast synthesis are very sensitive to water deficits. Opening of leaf pores and assimilation (incorporation) of carbon dioxide are moderately sensitive to water stress. Respiration, conductance of water through the plant (xylem conductance), and sugar accumulation are generally less sensitive to water stress than the other plant processes mentioned. The primary, direct effect of drought stress is cell dehydration, which is completely reversible up to a point.

Water deficits can also damage peanut through a variety of indirect stress-related injuries, including phosphorus and calcium deficiencies, reduced growth resulting from an impaired ability to translocate, reduced root nodule formation and function, and metabolic disturbances such as reduced photosynthesis and increased respiration. The following sections discuss the physiological basis for the various types of damage related to water stress in peanut.

Plant Turgor. Reduced plant turgor is one of the first effects of a water deficit. A decrease in plant turgor will slow new plant growth, inhibit translocation, and decrease flowering and peg and pod elongation. Expansive growth is very sensitive to even slight tissue water deficits.

Indirect Nutritional Effects. Indirect nutritional effects are thought to result from impairment of translocation in water-deficient plants. An accumulation of photosynthetic products may occur in the leaves because of loss of plant turgor and a resulting delay in the transfer of these products from the photosynthetic tissue to the conducting tissue.

Nutrient Deficiencies. Phosphorus and calcium deficiencies can develop as a result of either impaired nutrient uptake or increased leakage from cells during drought. Adequate calcium is important to cell wall function (Chapter 2). Because of the effect of drought on seed calcium concentrations, seed harvested from drought-stressed peanut plants are prone to germination problems.

Nodule Formation and Function. Impaired nodule formation and function can result from low soil moisture during seedling growth. With severe water deficits, nodule formation can be inhibited, even when a *Bradyrhizobium* inoculant has been applied. Late-season drought can cause cells on the nodule surface to collapse, decreasing the exchange of air (nitrogen and oxygen in particular) with the soil. Reduced nitrogen fixation may contribute to the chlorotic look that drought-stressed peanut plants have during the stress as well as for several weeks after the stress is relieved.

Metabolism. Metabolic disturbances are associated with moderate to severe drought stress. Changes occur in the rate of respiration, the availability of energy carriers for cells (adenosine triphosphate [ATP]), and the stability of cell membranes as the plant enters a severe water deficit. Increases in enzyme system activity (for example, ribonuclease activity) and the breakdown of carbohydrates (glycolysis) have been associated with drought. Biochemical lesions (deficiencies of intermediate building blocks) and the accumulation of possible toxic products, such as ammonia (NH_3) from protein breakdown, may develop as a result of decreased activity in some enzyme systems. Stresses caused by water deficits may cause changes in hormone production. Effects may include decreases in growth-promoting hormones, such as the cytokinins, and increases in growth inhibitors, such as abscisic acid.

Often plants will move soluble substances and water from lower to upper leaves during the onset of wilting. Protein breakdown to amino acids is common in water-stressed plants. Increase in the amino acid proline is often associated with drought-stressed leaves. However, whether proline accumulation is an indirect result of protein breakdown or whether it plays a direct role in drought resistance is unknown.

Photosynthesis. Inhibition of photosynthesis occurs as a result of partial or complete closure of stomata, which is one of the early responses of water-stressed peanut to conserve water. This response, however, slows the movement of carbon dioxide into the leaves and sharply reduces the rate of photosynthesis.

Carbohydrate Reserves. A decrease in reserve carbohydrates occurs as a result of an increase in the ratio of respiration to photosynthesis, primarily because of a sharp decline in the rate of photosynthesis.

Pod Splitting. Pod splitting is common when water deficits are relieved suddenly. The long-distance transport of water and nutrients to the seed occurs via the phloem and is driven by a turgor pressure gradient between the leaves and the seed. When a plant is water deficient, turgor pressure will be low throughout the plant. Turgor pressure increases when the water deficit is released, causing an increased flow of water and nutrients to the seed.

Cell growth in the seed, as in the rest of the plant, is in part turgor driven. When moisture is limited during early pod development, pods will be small. The endocarp of the pod normally becomes rigid before the seed undergoes its rapid swell phase. If the water stress is relieved suddenly just as the seed is starting to swell, the seed may enlarge to the size it would have been in the absence of water stress. Since the seed is enclosed in the smaller-than-normal hull, a crack in the hull is likely to develop, exposing the seed to the soil. Pods in the yellow 1 and early yellow 2 stages (on the basis of the hull-scrape technique [Chapter 5]) are in the most danger of this physiological cracking disorder. Seeds already past this stage (late yellow 2 and beyond) are unlikely to swell further, since seed size is already fixed. Figure 6.3 illustrates the effect of moisture stress on seed size distribution. Note that the more severe the stress, the greater the percentage of small seed.

Soil-to-Root Water Transport. Impairment of soil-to-root water transport occurs as water becomes depleted from the soil. As the soil dries, both soil and root shrinkage occur, resulting in less root surface area to make contact with soil particles. As a result, the flow of water to the root may be impeded.

Xylem Water Transport. Breaks in the water columns within the xylem result in impairment of water transport. These breaks are caused by gases dissolved in the vascular water that come out of solution and form gas pockets (emboli) in xylem cells. Emboli form even at moderate (8–15 bars) soil water tensions. Vascular emboli gradually disappear as water is restored to the plant. However, repair is best achieved when the transpiration demand is low, such as under cloudy, rainy conditions or at night. The importance of the disruption of water columns within the xylem is not fully understood at this time, particularly in herbaceous plants.

The time during the growing season at which a water deficit occurs is a prime determinant of the peanut crop's ability to recover or withstand damage. The various effects of drought stress on peanut can be categorized according to its occurrence in early season (0–50 days after planting), midseason (50–100 days after planting), or late season (more than 100 days after planting). Midseason drought stress generally has the greatest effect on peanut yield. Peanut plants are quite resilient and can recover well from early season water deficits.

Midseason drought stress generally has the greatest effect on peanut yield.

As long as the peanut seed have adequate water for germination and crop establishment, early drought stress will have only a moderate effect on the yield and grade of the crop. Immediately after germination, the peanut seedling rapidly extends its taproot deep into the soil. The young seedling's tiny canopy is supported by a vast root system, which develops rapidly if it has sufficient water. This extensive root system equips the plant to withstand a moderate early season drought. However, water stress early in the season delays the time to maturity and harvest.

Water stress early in the season delays the time to maturity and harvest.

Water deficits at midseason will decrease the number of flowers as well as the number of flowers developing into pegs and the number of pegs developing into pods. In addition, drought will reduce the number of sound, mature kernels that are harvested and increase the number of aborted seeds and "pops."

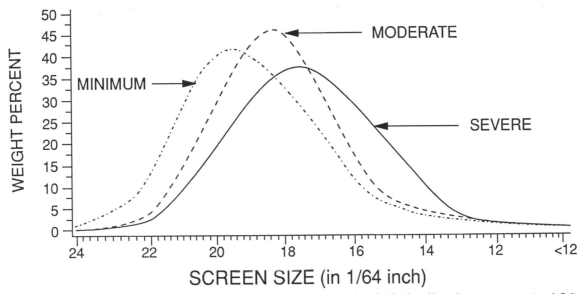

Fig. 6.3. Effects of minimum, moderate, and severe moisture stress on seed size distribution. Note that a severe water deficit results in a higher percentage of small seed. (Courtesy T. H. Sanders)

As with early season drought, late-season water stress typically does not affect yield as much as water stress during the critical midseason fruiting period. However, late-season drought is associated with a high incidence of aflatoxin contamination (Chapter 13). In addition, late-season water stress can increase the amount of pod loss caused by digging problems. The effects of water deficits at different times during the growing season on yields and grades of Florunner peanut are illustrated in Table 6.1.

Box 6.3 contains management tips for reducing the impact of drought.

Table 6.1. Effect of water stress at different times during the season on yield and grade of Florunner peanut

Treatment[a]	Yield (lbs/acre)	Grade (% SMK[b])
Well-watered control	4,470	75
Drought		
30–65 DAP	3,830	75
65–100 DAP	3,000	71
100–135 DAP	4,010	76

[a] DAP = days after planting, during which time the plants received no irrigation or rainfall.
[b] Sound, mature kernels.

Box 6.3

Tips for Reducing the Impact of Drought

- If irrigation is available, it should be used to relieve water stress. Generally, one application of 1.5 inches is preferable to two or more applications of smaller amounts.

- If water is available but limited (from an irrigation pond, for example), use it during the critical midseason fruiting period. Keep enough water available to protect the crop from an extended (30 days or more) late-season drought, which could cause peanuts to become contaminated with aflatoxin.

- Small soil dikes, placed periodically in the wheel tracks and in breaks between beds, can be used to help decrease water runoff.

- Providing conditions for establishing and maintaining a healthy root system aids in water uptake and drought resistance. Improving soil tilth by rotation with a grass crop, decreasing soil compaction, maintaining proper soil nutrient concentrations and subsoil moisture, and controlling soilborne insects and diseases all contribute to improved root function.

- Soil crusting can impede the penetration of water into the soil. Preventing or breaking up soil crusting will help water infiltrate the soil when it does become available.

- Knowledge of local rainfall patterns should be used to select planting dates most likely to avoid midseason drought.

- Row orientation and row spacing may affect moisture conservation. Research from Oklahoma suggests that peanut planted in a north-south orientation may conserve more water than rows in an east-west direction. Indications are that the best row orientation is perpendicular to wind direction.

Excess Water (Waterlogging)

A peanut crop can be subjected to stress from an excess of water if a heavy irrigation is followed by rain or if there is a poorly drained area in the field. Water absorption and transpiration are rapidly decreased by flood conditions. As with plants experiencing drought stress, plants in waterlogged soils show wilting leaves and stomatal closure. Alterations in root and stem morphology, formation of additional roots on the stems, leaf yellowing, and leaf loss are also associated with waterlogging (Plate 33). Excess water in the soil leaches nutrients out of the root zone and replaces most of the air in the soil with water. These conditions cause oxygen deficits, anaerobic respiration, and the reduction of both shoot and root growth.

Although plants are capable of anaerobic respiration, the toxic products that accumulate in conjunction with the lack of free oxygen can injure them. Since regeneration of the cells' energy-carrying molecules (ATP) comes mainly from aerobic respiration, energy for all the ATP-utilizing pathways will be limited when free oxygen is lacking. This restricts the plant to the breakdown of carbohydrates for ATP replacement.

Flooding of the root zone may initially result in a low-oxygen environment. Low oxygen concentrations in the soil will not always affect root respiration but will probably impair root extension. Poor soil aeration may also cause a buildup of carbon dioxide, which can inhibit root growth.

In wet soils, intercellular air transport to the root tips from the aboveground stem becomes important. This transport is aided in some plants, such as wetland rice, by a special cortical tissue known as the aerenchyma. The nonaerenchymous structure of peanut root cortical cells does not allow enough gas exchange to keep a large root system supplied with oxygen.

Ethylene, a gaseous hydrocarbon produced by roots and by microbial activity, is also known to accumulate in poorly aerated soils where water prevents the escape of the gas. In addition to restricting root growth, ethylene is known to inhibit photosynthesis in peanut leaves.

Ridges and lenticels, raised (usually white and puffy) spots on the root and nodule surfaces (Plate 34), can allow ethylene to escape from the root area. Reduced diffusion of gases to the root surface can inhibit nitrogen-fixing nodule formation and reduce the effectiveness of nodules already formed. Lenticels in waterlogged soil expand to aid in gas diffusion into and out of the nodule.

Reduced root growth and activity will decrease the absorption and translocation of nutrients to the shoot. Nutrient transport from the leaves to the roots will also decrease as a result of soil flooding. Some soil nutrients will change from the oxidized to the reduced state, changing their availability. For instance, iron oxide (Fe_3^+) will convert to the more soluble and toxic form, Fe_2^+. Toxins such as nitrites (the NO_2^- form) can also accumulate as a result of anaerobic microbial activity on nitrate nitrogen (NO_3). Even the activities of pathogens such as *Phytophthora* and *Pythium* spp. are enhanced in wet soils before oxygen depletion.

Tips for reducing the impact of excess water are summarized in Box 6.4.

Radiation Stresses

Radiation stresses include damage mainly from excessive or insufficient light and ultraviolet (UV) radiation. The most common is stress from insufficient light interception caused by weed competition or defoliation.

Excessive or Insufficient Light

Peanut leaf canopies can be thought of as loosely stacked groups of small leaflets. Light gaps or spaces between the leaflets allow radiation to penetrate the canopy. Light may also penetrate by reflectance, which depends on the leaf thickness and shininess, and by transmission, in which light passes through the leaf.

Dense plant populations generally have higher leaf area index values (square inches of leaf per square inch of ground) than less dense populations. Light intensities at the soil surface beneath a full peanut canopy of six to eight layers of leaves can be less than 5% of the intensity of unshaded areas. As a shaded leaf in the peanut undercanopy grows, it can adapt and improve its light-harvesting efficiency by reducing its respiration rate, by reducing its surface and subsurface wax and hair development to minimize reflectance from the leaf surface, and by increasing its size.

Light conditions also affect the shape of a peanut canopy. Under shaded conditions, a greater percentage of the absorbed

---

Box 6.4

Tips for Reducing the Impact of Water Excess

• To minimize the amount of water added to a wet area in the field, irrigation equipment can be modified to control the output of individual nozzles and the traveling speed of the system.

• Because no large field is completely uniform in its ability to hold moisture, application of water should be dependent upon actual soil moisture conditions throughout the field. Irrigation water should be managed by using soil moisture-sensing devices to help prevent problems associated with uneven distribution of moisture in a field.

• Cover crops extract moisture from the soil and can be used to help reduce moisture excesses in a field.

• Tile drainage is often used to remedy a wet area in a field if the area is not considered a wetland.

Box 6.5

Tips for Reducing Radiation Stresses

• Follow proper pest-control programs to maintain a healthy canopy and reduce light competition from weeds. Remember that young leaves are the most efficient light harvesters.

• Avoid water stress and nutrient deficiencies that will affect the photosynthetic capacity of the crop.

• Use vigorous seed, and uniformly space plants to maximize light gathering by the crop.

• Avoid treating the crop with compounds that reduce the protective waxy layers of the leaves (see section on chemical stresses).

---

nutrients are used for main stem growth than under unshaded conditions. For example, dense plant populations have plants with taller main stems and, to a lesser degree, longer, more upright lateral branches than plants in lower population densities where there is less competition for light (Plate 35). In general, peanuts grown in weedy fields will have longer stems than peanuts grown without weed competition.

Part of this elongation effect may be caused by shading, which changes both light quality (color spectrum) and quantity. High light intensities may change the hormonal balances in the plant and reduce vine elongation. However, experiments have shown that the effect of shaded conditions on peanut plant height relates more to a lower quality of light than to a lower total quantity of light. Plant height is affected more by low-intensity infrared light than by shorter days.

Because of physiological adaptations, shaded leaves are more efficient in harvesting light under low light conditions than are leaves developed in the sun. Shaded leaves adapt to high light intensities by increasing their development of waxy layers on leaf surfaces. However, when leaves grown under low light intensities are suddenly exposed to high light intensities, a secondary radiation-induced stress injury, called sun scald, can result.

Leaf age is an important factor in photosynthetic ability. The efficiency of photosynthesis increases from the time a peanut leaf unfolds until it is about 20 days old. From that time on, photosynthetic efficiency declines. When a leaf is 50 days old, it may be only 20% as efficient as it was at its peak. As leaves "wear out," the peanut plant constantly replaces them. Management practices to assist in maintaining a healthy canopy will help the plant retain its photosynthetic activity.

Interestingly, light plays an important role in pod formation as well as in plant and leaf morphology. The absence of light has been found to be an essential factor in pod formation. Pegs will continue to elongate until they reach a soil depth with little light penetration. For this reason, pod set is deeper in light, sandy soils than in heavy soils.

Ultraviolet Radiation

Ultraviolet radiation damage to plants is rare. However, UV radiation can destroy proteins and nucleic acids, inactivate enzymes, destroy organelles, and damage cell membranes. The effect of UV radiation on plant hormones, particularly auxin, is also thought to inhibit plant growth.

Plants are protected from harmful levels of UV radiation by various mechanisms, including the formation of epidermal waxes, cutin, and cell solutes. Plant pigments such as anthocyanin (a red pigment) absorb strongly in the UV range. This pigment is usually found in the sap of the outer cell layers of a leaf and is thought to provide some protection against damaging levels of UV radiation.

In general, plants have protective mechanisms against UV radiation with wavelengths below 300 nm, which are potentially the most damaging. Longer UV wavelengths, however, may injure plants. For example, the lighter green color of leaves exposed to direct sun may be caused in part by the destruction of chlorophyll by natural UV radiation.

In addition to producing a physical, protective barrier at the leaf's surface, plants have photoreactivating enzymes that can repair UV radiation damage to DNA and RNA. Certain amino acids, such as proline, are also thought to help the plant resist UV damage. The levels of these compounds found in leaves exposed to full sunlight are often higher than those found in shaded leaves.

Box 6.5 contains tips for reducing radiation stresses.

Chemical Stresses

Chemical stresses to peanut may result from plant exposure to fertilizers, herbicides, fungicides, insecticides, salts, or even gases such as ozone, sulfur dioxide, and ethylene. Chemical stresses on a crop can be reversible or nonreversible, primary or secondary. Like other stresses, peanut cultivars will differ in their ability to tolerate or avoid these stresses. A few key examples will be used to show how some chemicals cause physiological stresses to plants, and Box 6.6 contains tips for reducing these stresses.

> **Surfactants, crop oils, and inert ingredients (which may include some strong organic solvents) are often applied with the pesticide's active ingredient.**

Pesticides

Chemical pesticide applications to the crop plant involve more than the application of the active ingredient. Surfactants, crop oils, and inert ingredients (which may include some strong organic solvents) are also often applied with the pesticide's active ingredient. These solvents, crop oils, and surfactants aid in the distribution and retention of the pesticide, but they may also damage the cuticle, or waxy layers, on the surfaces of the leaves.

While pesticides can kill pests, preventing significant crop damage, they can also have negative effects on nontarget plants. An herbicide such as 4-(2,4-dichlorophenoxy)butyric acid (2,4-DB) (Chapter 7) has a growth-regulating effect when it is metabolized to 2,4-D. Peanut is less affected by 2,4-DB than are weeds such as cocklebur because peanut breaks down 2,4-DB to 2,4-D more slowly than cocklebur, preventing the peanut plant from being killed. However, peanut does suffer some negative effects of 2,4-D, such as stem twisting and seed deformation.

Salt

In many parts of the world where peanut is grown, evaporation exceeds precipitation, and both irrigation water and the soil are moderately saline. The amount of salt (particularly sodium chloride) a peanut crop can tolerate depends on the cultivar and the texture and water-holding capacity of the soil. High salt content in the pod zone can decrease pod and seed weight and cause dark areas on the inner faces of the cotyledons. High salinity in the rooting zone will decrease nitrogen fixation, decrease vegetative weight, and increase the percentage of immature pods. Other symptoms associated with salt injury include burning of the leaf margins, loss of turgor, loss of leaves, and, if the injury is severe enough, death of the plant. Salt stresses are often directly comparable to water stresses. Salt reduces the osmotic potential of water, thereby making it less available to the plant, and can induce a physiological drought stress on nontolerant plants.

When peanut cultivars that are not salt tolerant are grown in saline soils, leaves grow more slowly, probably because of decreased turgor pressure in the leaves and plant hormonal changes. Leaf photosynthesis and transpiration decrease, as do nutrient uptake and elimination of toxic diamines, which are formed only under salt-stress conditions. Nutrient absorption is changed as sodium chloride competes with needed nutrients such as calcium, magnesium, phosphorus, and potassium.

> **Symptoms associated with salt injury include burning of the leaf margins, loss of turgor, loss of leaves, and, if the injury is severe enough, death of the plant.**

To avoid salt stress, some salt-tolerant plants have roots that are able to prevent sodium or chlorine from entering the vascular tissue and being transported to the leaves. Some plants can also respond to salt stress by compartmentalizing the salt in older leaves and shedding them or by excreting excess salt from the leaves. As in other plants and in other stresses, it has been suggested that the stimulation of proline formation adds to the salt tolerance of peanut.

Nutrients

Peanut has nutritional requirements similar to those of legumes such as soybean and pea, with the exception of certain elements. Two of the more common nutrient stresses observed in peanut are calcium deficiency and zinc toxicity (Chapter 2).

Gaseous Stresses

Crop losses caused by air pollution are of increasing concern to some agriculturists. Perhaps the gaseous stress of most concern to agriculture is that caused by ozone. Recent estimates of ozone damage range from no effect to crop losses of up to 20%. While peanut is not as sensitive to ozone damage as soybean, it is more sensitive than barley.

Box 6.6

Tips for Reducing Chemical Stresses

- Tank mixing several pesticides may cause synergistic or antagonistic effects on the pest and the crop. Seek advice from extension service personnel or pesticide company representatives before trying new combinations.

- Maintain proper pH and soil nutrient levels.

- Avoid placing heavy doses of K or Mg fertilizers in the pod zone because these ions compete with calcium. High calcium levels in the pod zone strengthen the hull and seed and help prevent fungal damage.

- Under saline conditions, manage irrigation water to reduce evaporation at or above the soil surface (reducing salt deposition at the surface) and to increase the leaching of salts out of the pod zone. With adequate leaching and drainage, irrigation water with dissolved salts up to 1,500 parts per million can be used.

- Select peanut cultivars carefully to avoid chemical stresses.

Ozone is a highly reactive oxidizing compound that damages leaf chloroplasts and interferes with nitrogen metabolism. Ozone produces minute blisters followed by membrane damage, cell collapse, dehydration, and the silvering or bronzing of the affected spot.

As with responses to most stresses, peanut botanical types and even cultivars within a type differ in their responses to ozone. In general, Valencia-type peanuts have exhibited less leaf bronzing damage than Spanish-type peanuts exposed to the same ozone level. In addition, the peanut plant seems to be able to harden itself against ozone damage. Plants exposed to low ozone levels before exposure to higher levels show less damage than those with no prior ozone exposure.

Physical Stresses

Often overlooked are the physical stresses caused by equipment, wind, and soil. These stresses can reduce the yield and quality of a peanut crop.

Wind

Wind increases the evaporation rate of water. However, its effect on water loss from the leaf is mitigated by the plant's ability to control stomatal openings and adapt to its environment. Like drought, increased wind can change the number and size of stomata on new leaves, induce a water stress, desiccate leaves, increase respiration, and retard stem elongation.

Winds can also induce damage by "sandblasting" the peanut plant, especially early in the season before canopy closure. Soil particles carried by the wind can wound tender leaves and stems, providing entry for pathogens such as *Rhizoctonia*.

Soil Compaction

Growing plant roots seek the path of least resistance. One of the advantages of growing a grass crop before peanut is that grass will provide many root channels in the soil that can later be used by peanut roots.

> # Grass will provide many root channels in the soil that can later be used by peanut roots.

Compacted areas in the rooting zone will generally be avoided if the root has an alternative path. For example, most of the roots of plants grown next to a wheel track will form toward the center of the bed; very few will extend toward the compacted wheel track. Similarly, field traffic will change the fruiting pattern of peanuts (especially of the runner types): more pods will form toward the center of the bed than on the side toward the wheel track.

Damage to Seed

Abnormal root formation and poor germination can result from improper handling of peanut seed. Because of its exposed radicle (the root end of the peanut embryo that sticks out of one end of the seed), peanut seed is very susceptible to physical damage.

Many practices can damage seed, including combining pods at moisture levels that are too low or too high, picking with improper combine settings, dropping the pods into the wagon, curing the pods at temperatures that are too high, dropping the pods from great heights into a warehouse, moving pods with a front-end loader, bruising seed during the shelling and bagging operations, and dropping the bagged and treated seed onto the farmer's truck (Chapter 5).

The percentage of abnormal roots coming from combine-harvested and mechanically processed seed will be three to five times higher than that of hand-harvested and hand-processed seed (15–25% and 5–8%, respectively). Slower growth, lower plant populations, and lower yields have been documented for commercially prepared seed when compared with hand-processed seed. In some cases, the yield of individual plants with abnormal roots is only 50% that of plants with normal roots.

Seed that has been mishandled often loses its seed coat. These "baldhead" seed are from three to nine times less likely to germinate, and seedlings from baldhead seed are often stunted and abnormal. Baldheads resulting from curing at temperatures that are too high may be dead before they are bagged. Viable baldhead seed do not function as well as intact seed, primarily because the intact seed coat helps the two seed cotyledons stay together and provides both a physical and a chemical barrier to bacterial and fungal pathogens.

Box 6.7 has tips for reducing physical stresses.

Selected References

Ablett, G. R., Roy, R. C., and Tanner, J. W. 1981. Agronomic aspects of normal root formation in peanuts (*Arachis hypogaea* L.). Peanut Sci. 8:25-30.

Boote, K. J., Jones, J. W., Mishoe, J. W., and Berger, R. D. 1983. Coupling pests to crop growth simulators to predict yield reductions. Phytopathology 73:1581-1587.

Cherry, J. H. 1989. Environmental Stress in Plants: Biochemical and Physiological Mechanisms. Ser. G, Ecological Sciences, vol. 19. Springer-Verlag, Berlin.

Ensing, J., Hostra, G., and Adomait, E. J. 1986. The use of cultivar yield data to estimate losses due to ozone in peanut. Can. J. Plant Sci. 66:511-520.

Fitter, A. H., and Hay, R. K. M. 1987. Environmental Physiology of Plants, 2nd ed. Academic Press, San Diego, CA.

Fry, W. E. 1982. Principles of Plant Disease Management. Academic Press, New York.

Jackson, M. B. 1985. Ethylene and responses of plants to soil waterlogging and submergence. Annu. Rev. Plant Physiol. 36:145-174.

Ketring, D. L. 1979. Physiology of oil seeds. VIII. Germination of peanut seeds exposed to subfreezing temperatures while drying in the windrow. Peanut Sci. 6:80-83.

Box 6.7

Tips for Reducing Physical Stresses

- Minimize traffic damage to the soil and the crop by establishing and maintaining the same traffic pattern every time equipment is moved through the field.

- Wind breaks can reduce wind speed and therefore reduce damage to the plant.

- Be very selective in choosing the seed supply. Make sure seed has high germination values and is free of baldheads. Treat seed as if they were eggs. Load them very carefully, don't stack them in a miniature mountain, and keep them cool and dry.

Levitt, J. 1980. Responses of Plants to Environmental Stresses. Vol. 2, Water, Radiation, Salt, and Other Stresses. Academic Press, New York.

Manners, J. G. 1982. Principles of Plant Physiology. Cambridge University Press, New York.

Marschner, H. 1986. Mineral Nutrition of Higher Plants. Academic Press, San Diego, CA.

Mozingo, R. W. 1981. Effect of cultivars and field traffic on the fruiting patterns of Virginia type peanuts. Peanut Sci. 8:103-105.

Pattee, H. E., and Young, C. T., eds. 1982. Peanut Science and Tech-nology. American Peanut Research and Education Society, Yoakum, TX.

Roberts, D. A., and Boothroyd, C. W. 1984. Fundamentals of Plant Pathology, 2nd ed. W. H. Freeman, New York.

Singleton, J. A., and Pattee, H. E. 1987. Effects of induced low-temperature stress on raw peanuts. J. Food Sci. 52:242-244.

Wilkins, M. B. 1987. Advanced Plant Physiology. John Wiley & Sons, New York.

Zaitlin, M., and Hull, R. 1987. Plant virus-host interactions. Annu. Rev. Plant Physiol. 38:291-315.

B. J. Brecke
Agricultural Research Center
University of Florida, Jay

CHAPTER SEVEN

Management of Weeds

There are many kinds of plants in the world, and all may have some useful place. However, some plants are not desirable in agricultural situations and may interfere with crop production. We call such plants weeds. In simple terms, a weed is a plant out of place. This definition may include any plant that reduces yield and quality of a crop, interferes with cultivation of desirable plants, has the potential for decreasing the value of plant products, or presents a hazard of poisoning man or animals. As it is with most row crops, weed management is an important part of peanut production.

Successful management of weeds is a challenge for those growing peanut. Weeds compete with the crop for nutrients, water, and light. In addition, these unwanted plants often reduce harvesting efficiency and may contaminate the harvested peanuts, thus reducing crop quality and value.

Weed control is a major expense in peanut production. An estimated $74.2 million was spent for the purchase of herbicides used in peanut production in the United States during 1991 (Table 7.1). Even with this large expenditure, weeds still caused an estimated loss of $56.2 million to U.S. peanut producers. A better understanding of weed management systems that utilize not only herbicides but also cultural and biological weed control methods would greatly help in reducing these losses.

Weed Problems in Peanut Production

Many different species of weeds infest peanut plantings in the United States. These include both annual and perennial species of grasses and broadleaves. The more common weeds found in peanut in the United States are illustrated in Plates 36–51. A survey conducted in 1992 listed the 10 most common (Table 7.2) and 10 most troublesome (Table 7.3) weeds in each state of the U.S. peanut-growing region. Nutsedges, morningglories, and crabgrass were listed among the 10 most common weeds in peanut in all nine of the states surveyed. Pigweed commonly occurs in seven of the peanut-growing states (Table 7.2). Weeds listed as being the most difficult to deal with in six states include nutsedges, Texas panicum, morningglories, and Florida beggarweed (Table 7.3). Nutsedges, morningglories, and Texas panicum are considered among the most common and most troublesome.

Peanut-Weed Interactions

Effects of Weeds on Peanut Production

Weeds can cause serious yield losses in peanut if they are not controlled. Mixtures of weeds such as large crabgrass and Florida pusley or large crabgrass and pigweed have caused yield reductions of up to 95%. Studies of individual weed species have indicated that season-long competition from some species can result in drastic yield losses. Cocklebur and fall panicum have reduced peanut yields by as much as 85%, while broadleaf signalgrass and sicklepod have caused losses of up to 70%. Other species, such as Florida beggarweed and horsenettle, are less competitive and may cause yield losses of 20–40%.

These yield loss values illustrate that the impact of weeds on peanut varies widely, depending on the species and density (plants per unit area). Therefore, it is difficult to generalize about the impact of weeds on peanuts. It is important to consider the individual weed species present when the potential losses from weeds are evaluated.

Threshold Levels

An economic threshold is reached when the expected loss caused by a pest exceeds the cost of control. The density of weeds in a normal cropping situation is often at a level that obviously exceeds the economic threshold. When only relatively low densities are present, usually after some initial weed management practice has been used, thresholds can be used in management decisions.

Table 7.1. Economic impact of weeds on peanuts in the United States during 1991 in millions of dollars[a]

State	Losses caused by weeds	Cost of herbicides
Alabama	10.0	9.9
Florida	3.8	4.3
Georgia	30.4	41.3
Mississippi	0.1	0.2
North Carolina	3.8	7.1
Oklahoma	4.7	4.4
South Carolina	0.3	0.4
Texas	2.4	6.3
Virginia	0.7	0.3
Total	56.2	74.2

[a] Source: Bridges, 1992.

43

Many economic thresholds have been developed for other pests, especially insects, but only a few weed species have been studied in enough detail for establishment of thresholds (Table 7.4). Full-season competition from fall panicum at a density of two weeds per 30 feet of row reduced peanut yield by 25%; broadleaf signalgrass caused a 7% yield loss; bristly starbur reduced yields by 15%; and horsenettle, Florida beggarweed, and sicklepod lowered peanut production by only 2%. This again illustrates the point that competitive ability varies widely among weed species and that it is important to consider both weed species and density when a decision about whether a control measure should be implemented is being made.

Critical Periods of Weed Competition

Because of its low, spreading growth habit, the peanut plant tends to be less competitive with weeds than many other agronomic crops. As a result, it is usually necessary to keep the peanut crop free of weeds farther into the growing season than is necessary for crops such as corn or soybean. Peanut needs to be kept free of the annual grasses fall panicum and broadleaf signalgrass for 6–7 weeks to prevent a yield loss (Table 7.5). Broadleaf weeds, such as Florida beggarweed and sicklepod, can reduce peanut yields if they are not kept out of the crop for at least 4–6 weeks. The minimum weed-free period for horsenettle is 6–8 weeks and for silverleaf nightshade, 12 weeks.

Weeds can be allowed to compete with the crop for a time without impacting yield but then *must* be removed to prevent a yield loss (Table 7.5). Fall panicum can be allowed to compete

with peanut for only 2 weeks before having a negative effect; silverleaf nightshade and broadleaf signalgrass have limits of 4 and 8 weeks, respectively. Florida beggarweed, sicklepod, and horsenettle can be allowed to compete with the crop for up to 10 weeks before they must be removed.

> **A weed-free period of 4–6 weeks is needed, and weed competition can be tolerated for only 6–8 weeks after emergence without a loss in crop yield.**

It is difficult to generalize about the critical period of interference since this critical period is species dependent. For most weed species studied thus far, it appears that the peanut plant needs a weed-free period to optimize yields. A weed-free period of 4–6 weeks is needed, and weed competition can be tolerated for only 6–8 weeks after emergence without a loss in crop yield.

Methods of Managing Weeds in Peanut

Successful control of weeds in peanut often requires the use of several different components in an overall management program. An integrated approach to weed control usually results in a high level of control at the lowest possible cost and with the least possible stress to the environment. A general summary of weed management methods is given in Box 7.1.

Table 7.2. Ten most common weeds in peanut[a]

Weed	Scientific name	Number of states reporting[b]
Nutsedge	Cyperus spp.	9
Morningglory	Ipomoea spp.	9
Pigweed	Amaranthus spp.	9
Crabgrass	Digitaria spp.	7
Texas panicum	Panicum texanum	6
Cocklebur	Xanthium strumarium	5
Florida beggarweed	Desmodium tortuosum	5
Prickly sida	Sida spinosa	5
Sicklepod	Cassia obtusifolia	4
Goosegrass	Eleusine indica	3

[a] Source: Dowler, 1992.
[b] Total of nine states reporting.

Table 7.3. Ten most troublesome weeds in peanut[a]

Weed	Scientific name	Number of states reporting[b]
Nutsedge	Cyperus spp.	9
Texas panicum	Panicum texanum	7
Morningglory	Ipomoea spp.	7
Pigweed	Amaranthus spp.	6
Florida beggarweed	Desmodium tortuosum	5
Sicklepod	Cassia obtusifolia	5
Eclipta	Eclipta prostrata	5
Prickly sida	Sida spinosa	4
Johnsongrass	Sorghum halepense	3
Croton	Croton spp.	3

[a] Source: Dowler, 1992.
[b] Total of nine states reporting.

Table 7.4. Peanut yield losses caused by various weeds[a]

Weed	Yield loss (%)
Fall panicum	25
Bristly starbur	15
Cocklebur	15
Broadleaf signalgrass	7
Horsenettle	2
Florida beggarweed	2
Sicklepod	2

[a] Two weeds per 30 feet of row were allowed to compete with peanuts for the entire growing season. Only one species was present for each yield loss determination.

Table 7.5. Critical periods of weed competition in peanut[a]

Weed	Period of weed competition exceeded	or	Weed-free period was less than
	Peanut yield was reduced when:		
Fall panicum	2 weeks		7 weeks
Broadleaf signalgrass	8 weeks		6 weeks
Florida beggarweed	10 weeks		4–6 weeks
Sicklepod	10 weeks		4–6 weeks
Horsenettle	10 weeks		6–8 weeks
Silverleaf nightshade	4 weeks		12 weeks

[a] Weeks after crop emergence.

Cultural Control

Preventing weeds from infesting peanut fields is often the most cost-effective method of weed management. Good field sanitation during the previous cropping season will prevent weeds from producing seed and reduce potential weed problems in peanut. Thorough cleaning of harvesting equipment prior to use can also help to prevent the movement of weed seed between fields.

Crop rotation is an often overlooked weed management tool. Some weeds are easier to control in crops other than peanut because of the availability of herbicide treatments that are not registered for use in peanut or because of the greater competitive ability of the rotational crop. For example, an infestation of Florida beggarweed can be controlled more easily when corn is used in rotation with peanut because herbicides that control this weed, such as atrazine, can be used in the corn. Rotating crops and herbicides will also help prevent development of herbicide resistance in weeds.

Managing the peanut crop to achieve optimum growth is another cultural practice that will aid in achieving the desired level of weed control (Chapters 2 and 3). A well-managed crop growing in fertile soil at the proper pH and with optimum control of disease and insect pests provides rapid canopy cover of the soil surface. This allows the crop to be more competitive, results in reduced weed growth, and lessens the need for either mechanical or chemical control measures.

Changes in other cultural practices can also enhance the competitiveness of peanut. Reducing the distance between rows has been shown to increase potential yield per unit area and reduce weed growth (Chapter 3). Both sicklepod and Florida beggarweed produce significantly less biomass when peanut is grown with narrow row spacings of 8–16 inches (20–41 centimeters). With the rows spaced close together, the period of weed-free maintenance required to achieve optimum yields is shortened. Thus, an increased level of crop competitiveness is achieved with no additional control.

> **An integrated approach to weed control usually results in a high level of control at the lowest possible cost and with the least possible stress to the environment.**

Proper fertilizer placement can reduce the competitive effects of weeds. Placing at least a portion of the fertilizer in a band under the row will allow the crop to benefit from the added fertility while making the nutrients less available to weeds growing between the peanut rows. However, this method of fertilizer placement is rarely used (Chapter 2).

It is generally recommended that fertilizer be broadcast and incorporated with a disk before the land is turned with a moldboard plow. Herbicide can then be applied and incorporated prior to planting. However, growers often prefer the convenience of turning the land and then broadcasting and disking granular fertilizer that has been impregnated with preplant herbicide(s). For both practices, it is important to ensure that the herbicide or fertilizer-herbicide combination is uniformly distributed and thoroughly mixed with the soil.

> **Preventing weeds from infesting peanut fields is often the most cost-effective method of weed management.**

Proper management of irrigation can help to give the peanut crop a competitive advantage. Watering only when soil moisture reaches a level that will not support optimum crop growth will provide maximum benefit to the crop and minimum benefit to weeds, especially those with shallow root systems.

Mechanical Control

Cultivation is a very important component of an integrated weed management program. Mechanical weed removal can provide a substantial increase in the level of weed control over that obtained with the use of herbicides alone. Improper cultivation, however, can lead to substantial increases in the incidence of soilborne diseases. An increase in peanut stem rot has been shown to correspond with an increase in the amount of soil covering peanut leaves and stems (Chapter 11). Prevent-

---

Box 7.1

Methods of Managing Weeds in Peanut

Cultural
- Management of the crop for optimum growth rate with proper fertilizer placement and use of supplemental water enhances the competitiveness of the crop at the expense of weeds.
- Narrow row spacing achieves more rapid canopy closure.

Mechanical
- Cultivation can provide effective early season control, but if done improperly (with too much soil movement that covers crop leaves and stems), cultivation can lead to serious disease problems.
- Hand weeding may be done on a small scale, but it is labor intensive.

Chemical
- Herbicides can provide cost-effective control.
- Herbicides remain an important component of an effective weed management system.
- There are three types of herbicide application:
  1. Preplant incorporated for annual grasses, small-seeded broadleaf weeds, and nutsedge.
  2. Preemergence for annual grasses, small-seeded broadleaf weed, and nutsedge.
  3. Ground cracking and postemergence for annual and perennial grasses, small- and large-seed broadleaf weeds, and nutsedge.

ing soil from covering peanut foliage reduces the potential for disease problems.

A weed control system that utilizes both herbicides and timely, careful cultivation provides the best weed control and highest level of economic returns. The key to such a system is timely and judicious use of both chemical and mechanical methods of controlling weeds.

Biological Control

The use of insects or plant pathogens to manage weeds is one facet of biological control (biocontrol). Agents used for biocontrol can be organisms such as insects, fungi, or nematodes. Some agents when applied in a field become established and self-sustaining. Others offer short-term activity and must be reapplied to obtain control over a longer period of time.

Potential biological control agents have been identified for several weed species that are problems in peanut (Table 7.6). An insect and a rust fungus appear to be effective against nutsedge. Pathogens have also been isolated that provide control of sicklepod and Florida beggarweed. A parasitic nematode has shown promise for control of silverleaf nightshade. Other species that are being evaluated for susceptibility to biocontrol include goosegrass, cocklebur, pigweed, and morningglories.

Most of these biocontrol agents, however, have some specific requirements, such as long periods of dew, or limitations, such as relatively short shelf lives, that are difficult to deal with under field conditions. As a result, no biocontrol organisms have thus far been marketed for weed control in peanut. Research is being conducted to develop techniques that overcome these difficulties, and it is hoped that biocontrol agents will be available for use in peanut in the future.

Chemical Control

Herbicides have been used in peanut for more than 40 years and remain a key component of an effective and economical weed management system. The use of these agrochemicals was quickly accepted by peanut growers because herbicides provide an attractive alternative to methods such as mechanical cultivation and hand weeding that are often less effective and more time consuming and expensive.

In order to achieve an acceptable level of weed control in peanut, it is usually necessary to use multiple applications of herbicides. These applications must be properly timed to realize the maximum benefit. An herbicide program designed to control both annual grass and broadleaf species often includes a preplant incorporated treatment followed by an application timed to coincide with peanut emergence, usually referred to as ground cracking, and by one or more postemergence, over-the-top applications. The severity of the weed problem encountered will dictate the number of applications actually required. Suppose, for example, that there are annual grasses and heavy infestations of sicklepod and Florida beg-

Table 7.6. Biocontrol agents being investigated for potential weed control in peanut

Agent	Type	Target weed
Bactra ventura	Insect	Purple nutsedge
Puccinia canaliculata	Fungus	Yellow nutsedge
Alternaria cassiae	Fungus	Sicklepod
Pseudocercospora nigricans	Fungus	Sicklepod
Colletotrichum truncatum	Fungus	Florida beggarweed
Orrina phyllobia	Nematode	Silverleaf nightshade

garweed in a field planted with peanut. The preplant herbicide application will probably suffice for the annual grasses. The at-cracking treatment will kill many of the sicklepod and beggarweed that emerge with the peanut plants, but a postemergence application of an appropriate herbicide will be needed 2–4 weeks later for adequate control.

When a preplant or preemergence herbicide treatment is being considered, some knowledge of potential weed problems is needed to make the appropriate choice. This can best be accomplished through the use of field records, which show the weed species present and the severity of infestation observed during the previous season. The best herbicide to control the anticipated weed problems can then be selected from the list of those registered for use in peanut.

Types of Herbicides

Preplant Incorporated. Preplant incorporated herbicides are normally used to control annual grasses, small-seeded broadleaf weeds, and nutsedge. They are usually not effective for control of large-seeded broadleaf species. In order to be effective, these herbicides must be thoroughly and uniformly mixed with the soil. Two of the most effective implements for incorporation are the power-driven tiller and the tandem disk. Thorough incorporation of herbicides with a tandem disk requires two passes over the field. The second pass must be made in a direction perpendicular to the first.

> **The at-cracking herbicide applications are often the most important part of an effective weed-control program because they provide early season control of many of the most troublesome species.**

Preemergence. Herbicides in this category are applied after the crop is planted but before seedlings emerge. Treatments applied preemergence are effective in controlling most annual grasses and small-seeded broadleaf weeds and nutsedge. These herbicides require rainfall or irrigation for activation.

Ground Cracking and Postemergence. "Ground cracking" describes herbicide application from the time peanuts emerge until approximately 2 weeks after emergence. Since many of the herbicides applied at cracking must contact weed foliage to be effective, the exact timing of application is often based more on weed emergence than on the stage of crop growth. Within the at-cracking timing window, application needs to be delayed until weeds emerge but must be made before they are too large to be effectively controlled.

The at-cracking herbicide applications are often the most important part of an effective weed-control program because they provide early season control of many of the most troublesome species. Herbicides applied at cracking provide control of annual grasses, small-seeded and large-seeded broadleaf weeds, and nutsedge. Some at-cracking herbicides cause substantial injury to the peanut foliage. However, the crop usually recovers rapidly from this initial foliar burn, and there is no impact on peanut yield. Herbicides are also registered for later season postemergence, over-the-top application for the control of perennial grasses and both annual broadleaf weeds.

Postemergence treatments allow the choice of herbicides to be based on observed rather than on anticipated weed problems. With development of more effective postemergence, over-the-top treatments and as computer decision aids become available, herbicides can be utilized more on an as-needed basis than is now possible. Weed management in peanut, in some instances, can be accomplished with only postemergence, over-the-top herbicides and timely mechanical removal.

Herbicide Application

Application of herbicides is similar in some respects but quite different in others from application of insecticides or fungicides (Chapter 14). The equipment used for application and the formulations of the products available are identical for many of these pesticides. There are differences, however, in some of the timings of application and in the placement of the pesticide being applied (Box 7.2). Fungicides and insecticides are usually applied in a band over the row or in the seed furrow at planting and not as broadcast treatments prior to planting of the crop. Herbicides, however, are often applied before the crop is planted, and thus broadcast application is quite common. Herbicides in granular form applied at planting are placed not in the seed furrow but as a band over the row. In general, most pesticides are applied with a sprayer or granular applicator. When applied as granules, the material is

applied as it comes from the package without amendment. Liquid formulations are usually diluted with water prior to application.

The most widely used method of applying herbicides is spraying (Chapter 14). The basic design of a typical sprayer has not changed for many years and consists of a tank, a pump, a pressure regulator, and spray nozzles. The area covered by the sprayer can vary. Applying herbicides in a band over the row reduces the total amount of herbicide applied per planted acre and thereby reduces the cost of the chemicals. A portion of the ground between the rows is left untreated, however, and will require cultivation to achieve an adequate level of weed control. Broadcast applications, on the other hand, require more herbicide per acre but often eliminate the need for mechanical cultivation.

Several different types of nozzles are available for the application of herbicides (Box 7.2). Flooding pressure nozzles are often used for broadcast preplant or preemergence applications where spray coverage of foliage is not a factor and where reduction of spray drift is important. Even-fan pressure nozzles are used for preemergence or postemergence banding of over-the-row treatments. With even-fan pressure nozzles, the edges of the spray pattern are not tapered, so spray volume is the same over the entire spray width. Flat-fan pressure nozzles are used for broadcast application of preplant, preemergence, and postemergence treatments. The spray pattern is designed to allow for overlap of spray from adjoining nozzles.

Granular application offers the advantage of eliminating the need for mixing the herbicide with water prior to application and therefore the need to transport large volumes of diluent to the field. Granular application of herbicides is used most often to apply a band directly over the row at the time of planting. The granular applicator is usually attached to the planter so that planting and herbicide application are accomplished in a single operation. Occasionally, a granular applicator is attached to a disk or some other tillage implement so that a preplant incorporated herbicide can be applied as a broadcast treatment. One difficulty encountered by those interested in granular application is that only a limited number of herbicides are available in a granular formulation.

Herbicides can also be applied through irrigation systems, a method termed chemigation (Chapter 14). Before an irrigation system can be used for herbicide application, it must be fitted with a series of check and safety valves to eliminate any possibility of contaminating the water source in the event of equipment malfunction. Sprinkler heads also need to be checked to ensure uniform water distribution.

Using an irrigation system to apply herbicides can be very cost effective. This is especially true for some herbicides that normally would require incorporation. These can be applied without subsequent tillage because of the large volumes of water applied through the irrigation system with the chemical. Care must be taken, however, to prevent water source contamination.

Symptoms of Herbicide Injury

Peanut plants can be injured by herbicides as a result of excessive application rates, improper or inaccurate application, improper timing of application, drift from adjacent areas of either spray droplets or chemical vapors, carryover from a previous crop, or unusual soil or climatic conditions (Table 7.7, Plates 52–58). Most herbicides registered for use on peanut will not cause excessive, yield-reducing damage if label directions are followed.

Box 7.2

Herbicide Application Methods

Spray
- Banding over the row reduces the total amount of herbicide used but often requires cultivation of untreated areas.
- Broadcast application requires more herbicide but often eliminates the need for mechanical weed control.
- Spray nozzles include the flooding pressure type for broadcast preplant or preemergence application, the even-fan pressure type for banded, over-the-row preemergence or postemergence treatments, and the flat-fan pressure type for broadcast application of preplant, preemergence, and postemergence herbicides.

Granules
- Eliminates the need to mix herbicide with diluent prior to application.
- Usually used to apply herbicide in a band over the row.
- May be used for broadcast treatment.
- Can be utilized preplant or preemergence.
- Only a limited number of herbicides are available in a granular formulation.

Chemigation
- Works best with a sprinkler irrigation system.
- Requires a system of check and safety valves to prevent water source contamination.
- Some herbicide labels prohibit this type of application.

If a situation is encountered in which herbicide injury is suspected, patterns of injury may aid in diagnosing the problem. For example, injury limited to the edge of a field might indicate chemical drifts. Streaks of injured plants through the field might be the result of sprayer overlap or a single nozzle with an excessive flow rate. Sometimes injury is noted at the ends of rows where an operator may have slowed the tractor preparing to turn, resulting in an excessive dose in that area. Improper rate of application or herbicide carryover, on the other hand, would tend to be more uniform over the entire field. When diagnosis of an injury problem is attempted, records of previous crops and herbicide use and of chemicals applied to the peanuts can be invaluable. Such records can help eliminate certain possible causes of the problem while indicating the more likely culprit or culprits. It is also important to keep in mind that certain nutritional and physiological problems as well as damage from diseases, insects, and nematodes may mimic injury caused by herbicides. These factors need to be considered before a final determination is made on the cause of the observed crop damage.

Integrated Weed Management

Integrated weed management involves the use of a systems approach to weed control (Box 7.3). This includes the use of a well-adapted peanut cultivar (Chapter 4) planted at the recommended seeding rate (Chapter 3) in rows spaced as close together as practical. It also includes use of the proper rate and placement of fertilizer in order to enhance crop competitiveness and minimize weed growth. Such a system also involves

timely, careful cultivation and judicious use of the most effective herbicides for the weed problems encountered. The use of preventive measures such as field sanitation, crop rotation, and proper harvesting technique to minimize the spread of weed seed is also important. By using such an

---

Box 7.3

Methods Used in an Integrated Approach to Weed Management

Cultural
- Good field sanitation in the previous crop
- Crop rotation
- Well-adapted cultivars
- Recommended plant populations and optimum row spacing
- Proper rate and placement of fertilizer
- Cultivation in a timely and careful fashion

Chemical
- Herbicide application at the proper time and rate
- Herbicide program designed to match the identified weed problem

---

Table 7.7. Types of injury that may occur to peanut from use of certain herbicides

Family	Trade name	Injury symptoms
Amide	Alanap	Inhibition of root growth; malformed shoots; altered geotropism resulting in roots growing upward
Benzoic	Banvel	Bending, twisting, and curling of leaves and stems; splitting of stems
Bipyridilium	Starfire (Paraquat)	Wilting and rapid desiccation followed by necrosis of treated foliage
Chloroacetamide	Lasso, Dual	Delayed seedling emergence; growth inhibition; crinkling or cupping of leaves
Dinitroaniline	Balan, Prowl, Sonalan, Treflan, Trifluralin	Inhibition of root growth, especially of secondary roots, which may appear short and stubby or swelled; plants wilt readily under moisture stress
Diphenyl ether	Blazer, Cobra	Inhibition of shoot growth; bronze, necrotic spots on leaflets; also may be necrotic areas on stems
Imidazolinone	Pursuit, Scepter	Cessation of shoot growth; shortening of internodes; chlorosis followed by death of leaflets
Phenoxy	2,4-DB, 2,4-D	Abnormal plant growth with stem leaf cupping. 2,4-DB often causes chlorosis at the base of peanut leaflet and downward cupping, resulting in an elongated appearance
Phenylurea	Cotoran, Fluometuron, Karmex	Interveinal chlorosis followed by leaflet death, which proceeds inward from the tip and margins
Sulfonylurea	Classic, Ally, Harmony Extra, Accent	Cessation of growth; shortened internodes; yellowing and chlorosis followed by death of leaf and stem tissue
Thiocarbamate	Vernam (Vernolate)	Delayed seedling emergence; seedling leaves may be cupped or crinkled; leaflets may be sealed at the margins
Triazine	Aatrex, Caparol, Bladex, Lexone, Sencor, Princep	Interveinal chlorosis followed by leaflet death, which proceeds inward from the tip and margins
Miscellaneous	Basagran	Mottling or speckling of leaflets with occasional yellowing or chlorosis
	Roundup, Ignite	Cessation of growth; wilting and chlorosis followed by necrosis (usually a slow process that may take 10–14 days)
	Command, Zorial	Leaflets bleached white; complete necrosis from lethal doses

approach, weeds can be controlled at the lowest possible cost in terms of both dollars spent and stress to the environment.

Managing Weeds in Reduced-Tillage Systems

Crop production with reduced tillage is becoming more popular throughout the United States. Lower inputs in terms of fuel, labor, and equipment are some advantages of reduced tillage. Even more important is the reduction in soil erosion achieved with less tillage. Controlling weeds, however, remains a major obstacle to successful conservation tillage in crop production.

A minimum-tillage production system is often thought to require an herbicide treatment to kill existing vegetation followed by a preemergence treatment to provide residual control. Later postemergence applications of herbicides may be needed to control weeds that escape the earlier treatment. In peanut, a moderately intensive herbicide program with multi-

ple applications is required to obtain adequate weed control. Additional research is needed to better define weed management systems for reduced tillage peanut production.

Selected References

Bridges, D. C. 1992. Economic losses caused by weeds. Proc. South. Weed Sci. Soc. 45:382-391.

Buchanan, G. A., Murray, D. S., and Hauser, E. W. 1982. Weeds and their control in peanuts. Pages 206-249 in: Peanut Science and Technology. H. E. Pattee and C. T. Young, eds. American Peanut Research and Education Society, Yoakum, TX.

Dowler, C. C. 1992. Weed survey—Southern states. Proc. South. Weed Sci. Soc. 45:397-399.

McWhorter, C. G., and Gebhardt, M. R., eds. 1987. Methods of Applying Herbicides. Monogr. 4. Weed Science Society of America, Champaign, IL.

Skroch, W. A., and Sheets, T. J., eds. 1978. Herbicide Injury Symptoms and Diagnosis. Cooperative Extension Service, North Carolina State University, Raleigh.

Joe E. Funderburk
North Florida Research and Education Center
University of Florida, Quincy

Rick L. Brandenburg
Department of Entomology
North Carolina State University, Raleigh

CHAPTER EIGHT

Management of Insects and Other Arthropods in Peanut

Peanut cultivars grown in each of the major peanut-producing regions of the United States frequently differ in harvest maturity and market type (Chapter 4). Despite major differences in production practices, some arthropod pests (e.g., mites and insects) are economically damaging in all regions. At the same time, some pests are serious economic constraints only in certain regions. There are numerous occasional pests of peanut, but only a few pose persistent economic threats. Producers must be concerned with both foliage- and soil-inhabiting arthropod pests, since economically important species occur in both habitats. Injury from arthropod pests can occur during any peanut growth stage, but the greatest losses usually occur during the reproductive stages (Chapter 1).

Foliage-inhabiting pests of peanut include thrips, spider mites, bollworms, armyworms, cutworms, and the potato leafhopper. Other pests that are rarely of economic importance are the velvetbean caterpillar, the threecornered alfalfa hopper, the green cloverworm, the rednecked peanutworm, and the soybean looper. There are fewer soil-inhabiting pests of significant economic importance. However, these species represent a

Foliage-inhabiting pests of peanut include thrips, spider mites, bollworms, armyworms, cutworms, and the potato leafhopper.

great challenge to peanut producers. Southern corn rootworm, lesser cornstalk borer, and wireworms are major soil-inhabiting pests. Whitefringed beetles and white grubs are pests of only minor economic importance. Table 8.1 gives additional information about these arthropod pests of peanut.

The susceptibility of peanut to injury from arthropod pests varies with the crop growth stage. Therefore, the economic impact of these pests is associated with the physiological age of the crop. Many pests have wide host ranges and are highly mobile. Most pests reproduce rapidly, and multiple generations occur during the growing season (sometimes through an array of cultivated crops and native vegetation). Population size increases, and movement from other crops can influence the final pest status in peanut fields. Consequently, a knowledge of pest biology in surrounding crops can contribute to the development of pest management recommendations. An understanding of pest biology within a peanut field is needed to develop efficient ways to reduce crop losses. This chapter focuses only on aspects of biology relevant to effective management of arthropod pests in peanut.

Southern corn rootworm, lesser cornstalk borer, and wireworms are major soil-inhabiting pests.

The peanut crop is a suitable habitat for a wide range of beneficial arthropod species that aid in the suppression of pests. Beneficial insects, mites, and spiders form a complex fauna, and these natural control agents help reduce populations of peanut arthropod pests. However, the intense management activities typical in peanut production can reduce beneficial populations. Even the use of fungicides for peanut disease management can affect arthropod pest populations by reducing the important fungal diseases of some pest species. Of course, many insecticides directly reduce natural enemy populations, and only limited research has been conducted on ways to preserve the indigenous natural enemies in peanut.

Conservation of natural enemy populations is a primary consideration for integrated pest management (IPM) programs. Detailed information on pest mortality from individual species or combinations of natural enemies is not available for most production situations. There is no doubt that the combined activity of these natural enemies prevents many pests from reaching economically damaging levels and reduces the severity of outbreaks of other pest species.

Box 8.1

Some Natural Enemies of Peanut Arthropod Pests

Ground beetles (*Calosoma* spp.)
Bigeyed bugs (*Geocoris* spp.)
Damsel bugs (*Raduviolus* spp. and *Nabis* spp.)
Earwigs (*Labidura* spp.)
Red imported fire ant (*Solenopsis invicta*)
Spiders (Araneidae)
Parasitic wasps (Braconidae and Ichneumonidae)
Parasitic flies (Tachinidae)

A complete list of all beneficial organisms would be inappropriate for this book, but several common inhabitants of peanut fields are listed in Box 8.1. There are many general predators, including ground beetles, bigeyed bugs, damsel bugs, earwigs, the red imported fire ant, and spiders. Many peanut pests are hosts for parasitic wasps and parasitic flies. There are also diseases associated with peanut arthropod pests, the most common of which include the fungal diseases that infect spider mites, lesser cornstalk borer, velvetbean caterpillar, and many other pests. Virus and bacterial disease organisms of peanut arthropod pests are also important.

Integrated Pest Management Programs

Although pest pressure differs in form or intensity among various geographical regions, considerable effort is expended to reduce injury caused by arthropods. Through trial and error, it has been determined that it is neither possible nor feasible to eliminate all economic losses attributable to arthropod pests in peanut. The primary focus of IPM programs is to consistently minimize economic losses from arthropod pests. This is done by the careful use of ecologically compatible management tactics.

Table 8.1. Common arthropod communities in peanut

Pest[a]	Geographical regions where important[b]	Plant parts injured[c]	Overwintering stages	Generations per year	Management tactics
Thrips (*Frankliniella* spp.)	TX-OK, VA-C, SE	B, L	All stages	3–5 or more	Preventive or curative—biological, cultural, insecticidal
Armyworms (*Spodoptera* spp.)	TX-OK, SE	B, L, F, PG	Pupae	4	Curative—biological, insecticidal
Bollworms (*Helicoverpa zea*)	TX-OK, VA-C, SE	B, L, F, PG	Pupae	3 or more	Curative—biological, insecticidal
Velvetbean caterpillar (*Anticarsia gemmatalis*)	TX-OK, SE	L	No	2 or more	Curative—biological, insecticidal
Green cloverworm (*Plathypena scabra*)	TX-OK, VA-C, SE	L	VA-C, no; TX-OK and SE, all stages	2 or more	Curative—biological, insecticidal
Granulate cutworm (*Agrotis subterranea*)	TX-OK, VA-C, SE	L, F, PG	Adults, pupae	2 or more	Curative—biological, insecticidal
Rednecked peanutworm (*Stegasta bosqueella*)	TX-OK	B	?	3	Curative—biological, insecticidal
Soybean looper (*Pseudoplusia includens*)	TX-OK, SE	L	No	1 or more	Curative—biological, insecticidal
Spider mites (*Tetranychus* spp.)	TX-OK, SE, VA-C	B, L	Adults	Every 4–14 days	Curative or preventive—biological, cultural, miticidal
Potato leafhopper (*Empoasca fabae*)	TX-OK, SE, VA-C	S	VA-C, no; TX-OK and SE, all stages	2 or more	Curative—biological, insecticidal
Threecornered alfalfa hopper (*Spissistilus festinus*)	TX-OK, SE, VA-C	S	Adults, eggs	3 or more	Curative—biological, insecticidal
Southern corn rootworm (*Diabrotica undecimpunctata howardi*)	VA-C	PG, PD	Adults	3	Curative or preventive—biological, insecticidal, resistant varieties
Lesser cornstalk borer (*Elasmopalpus lignosellus*)	TX-OK, SE	L, PG, PD, S	All stages	4 or more	Curative or preventive—cultural, biological, insecticidal
Wireworms (*Conoderus* spp.)	TX-OK, VA-C, SE	PD	All stages	1 or fewer	Curative—insecticidal
Whitefringed beetles (*Graphognathus* spp.)	TX-OK, VA-C, SE	S, PD	Eggs, larvae	2	None
White grubs (Scarabaeidae)	TX-OK, VA-C, SE	PD	Eggs, larvae	1 or fewer	None

[a] The host range of the velvetbean caterpillar and green cloverworm includes primarily legumes; the lesser cornstalk borer prefers grasses and legumes but may be a pest on many families of plants; the other pests listed infest peanut and many other families of plants, but little is known about the host range of the rednecked peanutworm.
[b] TX-OK = Texas-Oklahoma production region; VA-C = Virginia-Carolina production region; and SE = Southeast production region.
[c] B = buds; PG = pegs; L = leaves; PD = pods; S = stems; and F = flowers.

Most peanut pests are detected by scouting during periods of risk. Scouts monitor crop growth stage, pest development and population density, and occasionally natural enemy development and population density. Management tactics for pest populations in individual fields are based on economic injury levels defined as the lowest population density of each pest that is likely to cause economic damage. The economic injury level usually changes during the growing season. It is a function of peanut growth stage and reflects changes in the response of the plant to injury from the pest.

The primary focus of IPM programs is to consistently minimize economic losses from arthropod pests.

Control measures should be applied only when pest density exceeds the economic threshold. This threshold is a pest density below the point at which control measures are needed to prevent an increasing pest population from reaching the economic injury level. In peanut, the most common control tactic is the use of insecticides. Effective, economical insecticides are available for most peanut pests. Efforts are made to use insecticides and rates that minimize harmful effects on beneficial natural enemies.

When employed in the management of arthropod pests, the economic injury level concept is usually simple and successful. The approach will not succeed if a pest's population density cannot be easily estimated in scouting programs. The economic threshold concept also fails when a fast, reliable method is not available to suppress a pest population that is approaching the economic injury level. The preventive use of insecticides for control of such arthropod pests is not usually a wise or recommended practice. Consequently, other control tactics are emphasized.

Alternative control tactics include host plant resistance, cultural control, and biological control. These are preventive control tactics employed as production practices without regard to specific pest densities within a field. Indigenous natural enemies in crop habitats serve as important biological control agents. Occasionally, beneficial species have been introduced by pest management specialists to control or reduce population levels of a pest.

Although information available for most peanut pests is substantial, there are important gaps. Available information on peanut response to pest injury consists almost exclusively of the effect of individual pest species on yield at specific crop growth stages. The physiological response of peanut to injury is poorly understood even for individual species and will be needed to begin managing injury guilds. An injury guild is a group of pest species that produce the same physiological plant response. Additionally, current ecological information is inadequate for the prediction of outbreaks of most pests. Therefore, IPM programs for peanut pests are based mainly on the detection and management of individual pest species.

Although decisions are generally made to manage individual pest species, the IPM concept ideally considers interactions of pest species with each other and the environment. Management decisions made for a single arthropod species can affect populations of other such pests. Although we are primarily concerned with management of arthropod pests in

this chapter, it is evident that management practices employed for some peanut arthropod pests may influence economic losses from weeds, diseases, or nematodes.

An injury guild is a group of pest species that produce the same physiological plant response.

IPM programs for peanut were designed for people making pest management decisions, namely, producers or their agents. Agricultural enterprises are small businesses that have their own sets of objectives, which may or may not coincide with objectives of the larger community. Operating procedures may involve decision making by producers, who are constrained by the wishes of the larger community. Therefore, we have government regulations. Pesticides are available for use at the discretion of individual growers but must be used according to label directions regulated by public agencies. The usage of land, water, and other resources is also regulated for conservation purposes, and sometimes these regulations impact IPM programs. Although the attention of IPM is directed toward optimizing resources, public concerns and policy are factors in the decision-making process. Overall, the adoption of the IPM approach has resulted in reduced chemical management of arthropod pests and enhanced environmental quality.

Management Programs for Major Pests

Arthropods that cause plant injury are classified into guilds, including leaf feeders, fruit feeders, stand reducers, assimilate sappers, turgor reducers, and architecture modifiers. Nearly all economically important peanut arthropod pests are leaf feeders or fruit feeders. Most recommendations in current IPM programs pertain to individual pest species. However, it is important to categorize the pests according to guilds, because the many pests within a guild cause the same type of injury. Injury, not insect populations, directly affects plant physiology. Disruption of physiological functions can decrease seed yield or quality. For example, armyworms, velvetbean caterpillars, soybean loopers, green cloverworms, bollworms, and others are leaf-mass consumers (Table 8.1). Producers should be concerned with the total amount of leaf feeding by all of these pests rather than the leaf damage caused by individual species.

Nearly all economically important peanut arthropod pests are leaf feeders or fruit feeders.

Other pests produce more than one type of plant injury; this increases the difficulty of understanding plant responses to injuries. For example, the lesser cornstalk borer sporadically reduces stands early in the season. Later in the season, the pest is primarily a fruit feeder, but it can also cause other types of injury. This and other aspects of pest biology complicate management recommendations.

A discussion follows of the management programs for the most economically important arthropod pests of peanut in the United States. This discussion includes relevant information on economic injury levels, sampling, and preventive and curative control tactics. It covers aspects of biology necessary for effective and efficient management. Problems associated with decision making and control are addressed.

Thrips and the Tomato Spotted Wilt Virus

Thrips are intracellular feeders that injure plants primarily by feeding on unopened leaflets (Plates 59 and 60). Numerous species inhabit peanut fields, but only three occur frequently: tobacco thrips, *Frankliniella fusca* (Hinds); western flower thrips, *F. occidentalis* (Pergande); and flower thrips, *F. tritici* (Fitch). Each species can inhabit peanut at numerous times during the growing season. However, the tobacco thrips is the principal species reaching high densities and causing plant injury. Several generations of this species typically develop through a growing season.

Environmental conditions during the spring favor population development and survival of thrips. The largest population densities occur in leaflet terminals during the seedling plant growth stages. Feeding causes distorted, scarred leaflets, which have reduced photosynthetic ability, and stunted young plants. Peanut plants can tolerate severe injury by thrips, and there is a considerable debate over the value of controlling thrips to prevent plant injury. Some studies report no pod yield losses resulting from thrips injury, but yield losses have been noted in other studies. Several investigators are studying the separate and interactive effects of thrips and herbicide injury on peanut seed, pod yield, and market quality. Preliminary results indicate that severe injury sometimes reduces pod yield and delays maturity. Consequently, managing thrips injury is now an important consideration in IPM programs. In the Virginia-Carolina production area, where cultivars with long maturity requirements are being produced in a relatively short growing season, preventing thrips from delaying crop maturity is crucial to profitable peanut production.

The curative approach is often recommended for management of thrips injury. Economic thresholds based on the percentage of damaged leaflets are used. Since peanut in all production areas can tolerate low to moderate levels of injury by thrips without economic loss, this is definitely a conservative and acceptable management option. There are several rapidly acting, effective insecticides labeled for thrips control in peanut.

Several granular insecticides also are efficacious against thrips and can easily be applied at planting. Most of the acreage in the Virginia-Carolina and Southeast production regions is treated in this preventive manner. These granular insecticides not only suppress thrips but also aid in controlling economically damaging nematodes. Studies indicate that insecticide performance varies among different peanut cultivars and possibly across geographical locations. Application of at-plant insecticides is not completely compatible with IPM philosophy, which emphasizes the avoidance of preventive insecticide applications. Little is known about the effects of insecticides used against thrips on nontarget pests and beneficial organisms. It is possible that preventive at-plant insecticides could increase the need for additional insecticide applications later in the growing season. There is also the potential for development of thrips populations resistant to intensively used insecticides. Alternating insecticides from different chemical classes can reduce the probability of resistance. For example, an insecticide of the carbamate class (aldicarb, oxamyl, or carbaryl) could be rotated with one from the organic phosphate (acephate, disulfoton, or fensulfothion) or synthetic pyrethroid (esfenvalerate or cyhalothrin) class.

Several peanut diseases in the United States are transmitted by insects. The only one that is currently a serious economic threat to peanut producers is tomato spotted wilt virus (TSWV), which is vectored by several species of thrips (Chapter 9). TSWV is the only major virus of peanut transmitted by thrips in the United States.

The major vectors in peanut are tobacco thrips and western flower thrips. TSWV has recently become established in peanut-producing areas of the United States, and epidemics of the disease have been noted recently in the Southwest and Southeast. The virus has a wide host range and is economically important in the production of numerous agronomic, vegetable, and ornamental crops.

> **TSWV is the only major virus of peanut transmitted by thrips in the United States.**

Primary infection, the introduction of the virus into a field, results from the movement of adult thrips to a host. After primary infection, secondary spread of the disease occurs when immature thrips feed on infected plants, mature, and pass the virus to uninfected plants within the field. Adult tobacco thrips and western flower thrips are both responsible for primary spread of TSWV in peanut fields, but secondary spread results from subsequent reproduction and dispersal of tobacco thrips.

Unpublished research by numerous investigators has revealed that initial infection of peanut is rarely prevented by controlling thrips with insecticides. Consequently, insecticides are not recommended in fields with low to moderate levels of TSWV. Insecticides reduce secondary spread where disease incidence is increasing rapidly.

Peanut can tolerate low to moderate levels of tomato spotted wilt disease without subsequent loss in seed yield or quality. The effect of disease incidence on seed yield and quality is poorly understood for most peanut growth stages. Disease incidence and crop tolerance, however, differ among commercial cultivars. When agronomically acceptable, the use of the cultivar Southern Runner may reduce the potential for losses from the disease.

Delayed planting in the spring reduces the level of thrips infestation and subsequent injury in peanut, but the influence on the incidence of tomato spotted wilt is not known. Cultural control will probably be important in managing thrips and TSWV in peanut in the United States. However, thrips biology and disease epidemiology must be better understood before cultural control strategies can be included in a management program that uses host plant resistance and insecticide control.

Lepidopterous Leaf-Feeding Guild

Lepidopterous (moths and butterflies) insect pests are those in which the injurious developmental stage is the caterpillar. Numerous lepidopterous species produce the same injury by eating peanut leaves. Damaging populations of fall army-

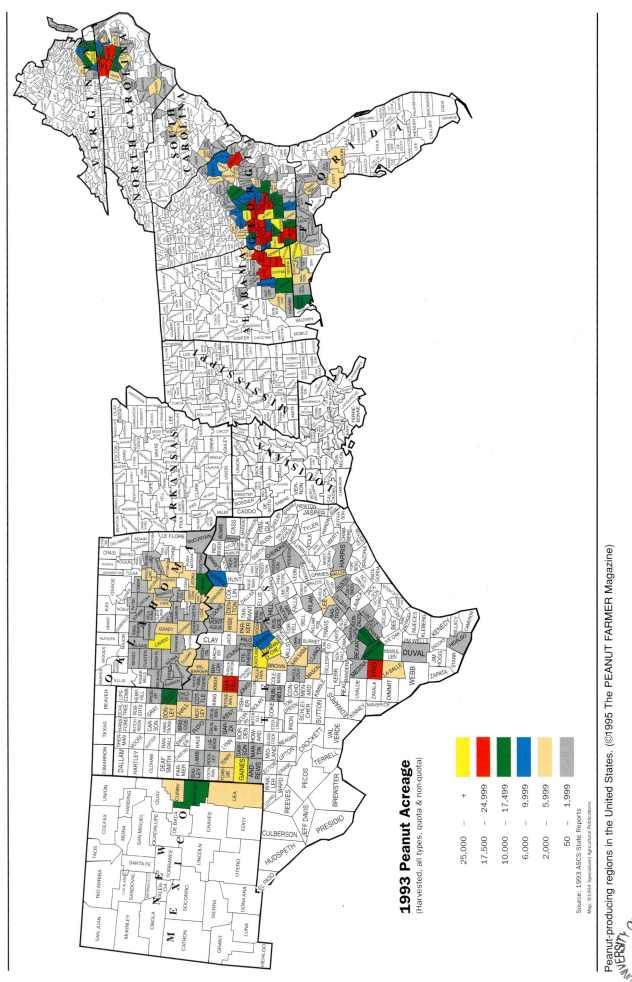

1993 Peanut Acreage
(Harvested, all types, quota & non-quota)

25,000	+
17,500 –	24,999
10,000 –	17,499
6,000 –	9,999
2,000 –	5,999
50 –	1,999

Source: 1993 ASCS State Reports
Map: ©1994 Specialized Agricultural Publications

Peanut-producing regions in the United States. (©1995 The PEANUT FARMER Magazine)

1. Yellow peanut flowers, a pointed peg, and a developing pod. (Courtesy H. A. Melouk)

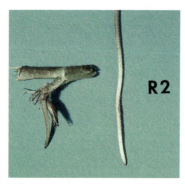

2. Beginning peg. (Courtesy K. Boote)

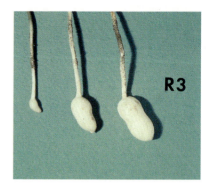

3. Beginning pod. (Courtesy K. Boote)

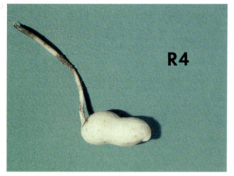

4. Full pod. (Courtesy K. Boote)

5. Beginning seed. (Courtesy K. Boote)

6. Full seed. (Courtesy K. Boote)

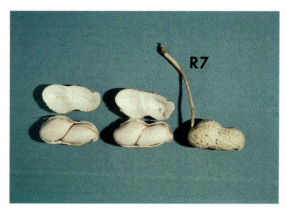

7. Beginning maturity. (Courtesy K. Boote)

8. Harvest maturity. (Courtesy K. Boote)

9. Overmature pod. (Courtesy K. Boote)

10. Symptoms of nitrogen deficiency. (Reprinted from Compendium of Peanut Diseases, American Phytopathological Society, 1984)

11. Symptoms of potassium deficiency. (Reprinted from Compendium of Peanut Diseases, American Phytopathological Society, 1984)

12. Symptoms of calcium deficiency. (Reprinted from Compendium of Peanut Diseases, American Phytopathological Society, 1984)

13. Symptoms of sulfur deficiency. (Reprinted from Compendium of Peanut Diseases, American Phytopathological Society, 1984)

14. Symptoms of iron chlorosis. (Reprinted from Compendium of Peanut Diseases, American Phytopathological Society, 1984)

15. Symptoms of manganese deficiency. (Reprinted from Compendium of Peanut Diseases, American Phytopathological Society, 1984)

16. Symptoms of copper deficiency. (Reprinted from Compendium of Peanut Diseases, American Phytopathological Society, 1984)

17. Leaves with marginal chlorosis caused by copper deficiency. (Reprinted from Compendium of Peanut Diseases, American Phytopathological Society, 1984)

18. Symptoms of boron toxicity. (Reprinted from Compendium of Peanut Diseases, American Phytopathological Society, 1984)

19. Chlorosis of leaves caused by zinc toxicity. (Reprinted from Compendium of Peanut Diseases, American Phytopathological Society, 1984)

20. Stunted growth caused by zinc toxicity. (Reprinted from Compendium of Peanut Diseases, American Phytopathological Society, 1984)

21. Stem splitting caused by zinc toxicity. (Reprinted from Compendium of Peanut Diseases, American Phytopathological Society, 1984).

22. Maturity composition of a crop of peanuts. Pod maturity varies from the most immature (left) to the most mature (right). A profile such as this is used with the hull-scrape method to determine the approximate digging date.

23. Digging peanuts with a digger-shaker-inverter.

24. Peanut plants inverted in windrows for drying.

25. Harvesting peanuts with a combine.

26. Combine ground speed must be set so that a continuous ribbon of plants flows into the machine. The pickup head should be adjusted so that fingers pick up the windrow without digging into the soil.

27. After combining, peanuts are dumped into a drying wagon of approximately 5-ton capacity.

28. Uninsulated steel warehouse used for storage of farmer's stock peanuts. Note the roof pitch with 33° slope. Natural ventilation is usually provided by large vents in the ridge and eaves.

29. Pile of farmer's stock peanuts inside a warehouse.

30. "Soldier," a large, molded clump of pods formed by condensation drip.

31. Pneumatic probe used to remove a sample of pods from a load of peanuts delivered to the sheller. The sample will be graded by the USDA Inspection Service.

32. Seed size distribution of Southern Runner peanut in relation to maturity. The numbers of large seed are greater in the mature peanuts.

33. Stunted plants with yellow leaves resulting from excess water.

34. Lenticels (white, puffy areas) on root of peanut plant grown in waterlogged soil.

35. Plants on the left have taller main stems than those on the right because of a higher within-row population. The plant population on the right has been reduced by stem rot. (Courtesy F. M. Shokes)

36. Purple nutsedge.

37. Pitted morningglory.

38. Ivy-leaf morningglory.

39. Pigweed.

40. Florida beggarweed.

41. Sicklepod.

42. Cocklebur.

43. Bristle starbur.

44. Giant horsenettle.

45. Tropic croton.

46. Prickly sida.

47. Broadleaf signalgrass.

48. Texas panicum.

49. Crabgrass.

50. Goosegrass.

51. Johnsongrass.

52. Paraquat injury on peanut.

53. Diphenyl ether injury.

54. 2,4-DB injury.

55. Trifluralin injury.

56. Triazine injury.

57. Vernolate injury.

58. Fluometuron injury.

59. Adult tobacco thrips, *Frankliniella fusca* (actual size about 0.1 inch).

60. Thrips injury to peanut leaves.

61. Corn earworm (*Helicoverpa zea*) larva and injury to peanut leaves.

62. Spider mite (*Tetranychus urticae*) colony with webbing on peanut terminal.

63. "Stippling" caused by spider mite feeding.

64. Spider mite damage (lower left) induced by insecticide.

65. Peanut pods damaged by southern corn rootworm.

66. Lesser cornstalk borer (*Elasmopalpus lignosellus*) larva with webbing on a peanut pod.

67. Symptoms caused by peanut mottle virus.

68. Symptoms caused by peanut stripe virus.

69. Symptoms caused by tomato spotted wilt virus on peanut. (Courtesy F. Killebrew)

70. Stunting of peanut plants caused by tomato spotted wilt virus. (Courtesy F. Killebrew)

71. Discolored, malformed peanut seed caused by tomato spotted wilt virus (left) and uninfected seed (right).

72. Leaf spot lesions on the upper side of a Florunner leaf.

73. Lesions of early leaf spot (*Cercospora arachidicola*) on the underside of a Florunner peanut leaf. Note the tan color that is characteristic of early leaf spot on Florunner and many susceptible cultivars.

74. Lesions of late leaf spot (*Cercosporidium personatum*) on the underside of a Florunner peanut leaf. Note the dark coloration. A few tan lesions of early leaf spot may be seen.

75. Tufts of olive brown spores (conidia) on two late leaf spot lesions (dark) next to an early leaf spot lesion (tan) on the underside of a peanut leaflet.

76. Aerial infrared photograph of peanut plants defoliated by late leaf spot. Dark areas on the left are plots of defoliated plants. Bright red plants have healthy foliage protected from leaf spot by fungicides (center) or are leaf spot-resistant breeding lines (right).

77. In-field, computer-controlled system for giving spray advisories for management of late leaf spot on peanut.

78. Pustules of the peanut rust fungus, *Puccinia arachidis*, on the underside of a peanut leaf.

79. Peanut rust damage on the cultivar Florunner (right), caused by *Puccinia arachidis*. Dead leaves tend to stay on plants. The rust-resistant cultivar Southern Runner (left) has very little damage.

80. Crown rot, caused by *Aspergillus niger*. Note the black masses of conidial heads on infected tissue. (Courtesy J. Damicone)

81. Germinating peanut seed infected with *Aspergillus niger*. (Courtesy H. Melouk)

82. Canker on peanut stem caused by *Rhizoctonia solani*. (Courtesy C. Lee)

83. Beginning *Rhizoctonia solani* lesion. (Courtesy F. Shokes)

84. Aerial blight caused by *Rhizoctonia solani*. (Courtesy F. Shokes)

85. Sheets of white mycelia of *Sclerotium rolfsii* on soil and infected stems. (Courtesy C. Lee)

86. Sclerotia of *Sclerotium rolfsii* on soil and infected plants. (Courtesy C. Lee)

87. Early symptoms of Cylindrocladium black rot. Note chlorosis (yellowing) and wilting of plants. (Courtesy B. Padgett)

88. Small, reddish orange perithecia (sexual structures) of *Cylindrocladium crotalariae*. (Courtesy B. Padgett)

89. Fluffy, cottony mycelia of *Sclerotinia minor* on branches near the soil line. (Courtesy H. Melouk)

90. Black sclerotia of *Sclerotinia minor* on the outside of an affected stem. (Courtesy H. Melouk)

91. Sclerotia of *Sclerotinia minor* between the cotyledons inside a pod. (Courtesy H. Melouk)

92. Shredded peanut stem caused by *Sclerotinia minor*. (Courtesy H. Melouk)

93. Yield reduction caused by *Sclerotinia minor* (right). The plant on the left is healthy. (Courtesy H. Melouk)

94. Symptoms of Verticillium wilt on peanut. (Courtesy H. Melouk)

95. Black discoloration of *Verticillium dahliae*-infected petioles (right) compared with healthy petioles (left). (Courtesy J. Damicone)

96. Pod rot caused by *Pythium myriotylum*. (Courtesy H. Melouk)

97. Blackened patches on surfaces of peanut pods caused by *Thielaviopsis basicola*. (Courtesy J. Damicone)

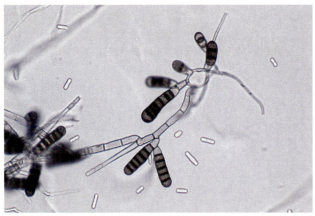

98. Black chlamydospores of the blackhull fungus, *Thielaviopsis basicola*. Note the presence of several hyaline conidia. (Courtesy J. Damicone)

99. Peanut plant infected with root-knot nematodes (right) compared with healthy plants. (Courtesy C. C. Russell)

100. Yellowing and stunting of peanut plants infected with root-knot nematodes. The middle green row was treated with a nematicide.

101. Galling, or knots, on roots caused by root-knot nematodes.

102. Extensive galling of peanut pods caused by root-knot nematodes.

103. Root proliferation of a peanut plant infected with northern root-knot nematodes. (Courtesy C. C. Russell)

104. Lesions on peanut pods caused by lesion nematodes.

105. Sporulation of *Aspergillus flavus* on a peanut kernel.

106. Sporulation of *Aspergillus flavus* on peanut pods.

107. Aflatoxin-contaminated peanut kernels.

108. Fumigation rig used to apply liquid fumigants. Note that the coulters in front of the chisels reduce the amount of trash that might accumulate. Fumigation tubes are mounted on the rear side of chisels, and fumigant is gravity fed through polypropylene tubing.

109. Typical applicators for use with granular pesticides. Granules are placed in the hopper box and metered through adjustable openings by an electric motor-driven rotor axle. Granules flow through tubes directly onto the plant or are spread out by baffles inside the fan-shaped row banders.

110. Aerially applying a pesticide to peanut.

111. Typical spray boom under the wing of an aircraft equipped for aerial pesticide application.

112. Micronair device used for ultra-low-volume spraying with aerial application.

113. Tractor-mounted hydraulic spray boom used to apply herbicides, fungicides, insecticides, and foliar-feed fertilizers.

worms, corn earworms (Plate 61), and beet armyworms sometimes occur during midseason or late in the season, while populations of velvetbean caterpillars sometimes occur late in the growing season. At all crop growth stages, peanut is tolerant of substantial levels of defoliation without economic damage. The amount of defoliation that can be tolerated varies with crop growth stage and environmental conditions.

Curative approaches, which easily allow integration of different control tactics, are recommended for the leaf feeders. All lepidopterous pests have natural enemies that frequently prevent populations from reaching economic threshold densities. When populations do reach outbreak densities, numerous labeled insecticides are available. The choice of an insecticide depends on the lepidopterous species in the field and the effect of the available insecticides on the natural enemies. For example, available *Bacillus thuringiensis* products provide control of most species and have little, if any, impact on beneficial populations. In addition, cultural control and host plant resistance can help reduce the possibility of outbreaks from lepidopterous pests. Planting early and/or selecting early maturing cultivars, for example, aids in the avoidance of velvetbean caterpillar infestations. Even when preventive practices are employed, fields should be scouted regularly.

> **Curative approaches, which easily allow integration of different control tactics, are recommended for the leaf feeders.**

Scouts use the shake cloth to collect caterpillars during most of the growing season. Population density is estimated by placing the shake cloth on the soil between two rows and vigorously shaking the foliage on each side onto the shake cloth. Care must be taken to shake each plant in the sample area. The foliage must be pushed back to allow quick identification and counting of all caterpillars. Late in the growing season when peanut plants are lapping, insects are shaken onto the soil, identified, and counted. The number of samples from a field varies according to field size.

> **Even when preventive practices are employed, fields should be scouted regularly.**

Although feeding potential varies with pest species and caterpillar developmental stages, economic thresholds are based on total numbers of caterpillars estimated (Box 8.2). Specific recommendations vary among states. For example, economic thresholds for Florida are three to four caterpillars per 30-cm (1-foot) row before the peanut plants meet between the rows, four to five caterpillars per 30-cm row after the plants meet between the rows, and five to six caterpillars per 30-cm row once plants have formed a complete canopy. These economic thresholds are very conservative, and unless

Box 8.2

Typical Scouting for and Managing of Lepidopterous Leaf-Feeding Insects

Ten random shake cloth samples were taken in each of two 20-acre peanut fields before mid-June to estimate densities of lepidopterous leaf-feeding pests. Lepidopterous leaf feeders were not economically important from mid-June to late August in these fields. Average densities of two corn earworm larvae, two fall armyworm larvae, and five velvetbean caterpillar larvae per row foot (30 cm) were noted in the first field. This exceeded the economic threshold density of five to six total caterpillars per row foot for peanut plants covering the row middles. In the second field, average densities of one corn earworm larva, zero fall armyworm larvae, and three velvetbean caterpillar larvae per row foot were observed. The field in which densities exceeded the economic threshold was treated the next day with a *Bacillus thuringiensis* product. This product is preferred because it suppresses each of the leaf-feeding pests while preserving natural enemy populations. Treatment was not needed on the second field. The populations of lepidopterous caterpillars were below the economic threshold density in both fields the following week and remained so for the remainder of the growing season.

extreme stress from drought or high temperatures occurs, peanut is tolerant of much larger population densities without economic damage. With further research, economic thresholds may be increased.

Spider Mites

Several species of spider mites inhabit peanut, but the only one that reaches economically important densities is the twospotted spider mite, *Tetranychus urticae* (Plate 62). This tiny pest of 0.12–0.16 inch (3 or 4 mm) damages plants by penetrating tissues and removing cell contents. Generally, the feeding is confined to the undersides of leaves. It causes the plant chloroplasts to disappear, and the leaves develop a speckled appearance called stippling (Plate 63). Photosynthetic processes in the leaves are disrupted.

Spider mite populations are usually greatly suppressed by many indigenous biological control agents, especially predators and pathogens. These natural enemies prevent spider mites from reaching economic threshold densities in peanut, but outbreaks sometimes are induced by application of certain pesticides (Plate 64). A pesticide may indirectly increase spider mite densities by suppressing indigenous biological control agents. For example, if a pesticide suppresses a fungal disease of spider mites, a rapid increase in the mite population could occur. Some pesticides may directly affect the pest's biology by increasing dispersal and reproduction and/or enhancing the suitability of peanut as a host.

Outbreaks of spider mites usually occur during periods of hot, dry weather. The reason for this is unclear, but apparently hot, dry weather favors spider mite reproduction and development while adversely affecting populations of natural enemies. Farming activities outside the peanut fields may also affect spider mite abundance within the peanut field. Spider mites have a wide host range and overwinter on weed hosts in field margins. These hosts and early spring crop hosts such as corn serve as nurseries for population growth and dispersal into peanut fields. Therefore, cultural control can be used to reduce the risk of spider mite outbreak. Since surrounding vegetation plays a critical role in introducing the pest to a peanut field, crop rotation away from corn reduces the amount of infestation. In addition, since mowing can cause dispersal into the field, field borders should not be mowed during the summer. It is important to avoid moving machinery from infested fields to uninfested fields. Irrigation is another cultural control tactic that reduces mite numbers during periods of hot, dry weather.

The tiny spider mites feed on the undersides of leaves, which makes scouting difficult. Usually the first indication of mite infestation is the appearance of large, discolored areas within the field. Such fields should be inspected frequently because of the high reproductive potential of spider mites. No economic thresholds for spider mites in peanut have been established. This lack of thresholds reflects the difficulty of scouting, the rapid reproductive potential of mites, and their sensitivity to pesticides and environmental conditions. Control is usually based upon a noticeable infestation, the grower's past experiences with the pest, forecasts for continued hot, dry weather, and the grower's concern about preventing further injury.

Box 8.3

An Insect-Resistant Peanut Cultivar

NC 6, a large-seeded peanut with a Virginia-type growth characteristic, was released in 1976. The cultivar possesses high levels of resistance to southern corn rootworm and moderate levels of resistance to thrips, potato leafhopper, and corn earworm. The release of NC 6 was a major step forward in the practical use of host plant resistance as a control tactic for management of arthropod pests in peanut. When it was released, agronomic characteristics of the cultivar were comparable to all other commercial cultivars grown in the Virginia-Carolina production region.

Resistance of NC 6 to southern corn rootworm is so high that one-fourth the normal rate of soil insecticide is recommended in poorly drained soils that are prone to severe infestation by this pest. For potato leafhopper and thrips, one-half the normal rate of insecticide is recommended. The overall reduction in the use of insecticides for these pests helps reduce the threat of problems later in the season from spider mites and corn earworm.

When the resistant cultivar was first released in 1976, it rapidly gained wide acceptance by peanut growers in the Virginia-Carolina production region. Recently, the percentage of acres planted to NC 6 has declined, primarily because new, higher yielding susceptible cultivars have been released. These new cultivars have good seed and shelling qualities, and growers expect a greater net economic return from growing the susceptible cultivar and using insecticides than from growing NC 6 with less insecticide or no insecticide. In addition, a cultivar that possesses disease resistance is more important economically than one with insect resistance. The cultivar NC 6 has demonstrated the benefits of arthropod-resistant cultivars to integrated pest management programs for peanut. It is hoped that cultivars with good agronomic characteristics and multiple pest resistance will be developed.

The only foliar pesticide available for controlling mite outbreaks in peanut is propargite. This product can be applied by air or ground, and good coverage of the plants, including under the leaves, is essential. A minimum of about 2 gallons of water per acre when it is applied by airplane and 8.5 gallons of water when applied by ground is recommended.

Nonchemical tactics should be emphasized in spider mite management programs.

Spider mites have a remarkable ability to develop resistance to pesticides. Therefore, nonchemical tactics should be emphasized in spider mite management programs. In addition to cultural control, management decisions for use of insecticides and fungicides should be carefully made to avoid creating or increasing spider mite problems. In most states, this information is incorporated into the control recommendations for other pests, such as caterpillars and leaf spot fungi. The cultivar NC 6 has moderate levels of resistance to spider mites and can be utilized to further reduce the possibility of economic problems (Box 8.3).

Southern Corn Rootworm

The most serious soil-inhabiting insect pest of peanut in the Virginia-Carolina production region is the southern corn rootworm (*Diabrotica undecimpunctata howardi*). Problems with this insect rarely occur in Southeast and Southwest production regions. Adults emerge from corn and other habitats during June or early July in the Virginia-Carolina region and are attracted to flowering peanut plants. Peak populations occur during late July or early August. Eggs laid by adults begin hatching during early August.

The larvae prefer to feed on immature, soft pods, but pegs may also be heavily damaged (Plate 65). Economic losses result from reductions in pod yield and market quality. Feeding also predisposes the pod to infection by soil microorganisms. Many factors influence the abundance of rootworm larvae in the soil, but the most important is soil moisture. Egg laying occurs only when soil moisture exceeds 5%, and few larvae survive after the first instar when soil moisture is below 2.5%. Eggs are usually laid near the base of the peanut plant where canopy shading preserves soil moisture.

The most serious soil-inhabiting insect pest of peanut in the Virginia-Carolina production region is the southern corn rootworm.

Soil moisture cannot be predicted accurately, but soil organic matter and clay content are used as general guides to the potential development of southern corn rootworm problems. In the Virginia-Carolina production region, preventive treatment with a soil insecticide is economically acceptable in any field with a soil organic matter content higher than 1.0%. If a field has a history of previous infestations, preventive treatment is also recommended.

The curative approach is frequently used in other production situations, even though scouting for this pest is laborious and technically difficult and granular insecticides require incorporation with water (either from rain or irrigation) to be efficacious. Curative approaches, however, can be used with moderate success if vigilant and frequent scouting is initiated when egg laying begins and if insecticide is applied immediately after attainment of economic thresholds. A commonly accepted economic threshold is either fresh damage or the presence of rootworms in 30% of the sites sampled. Scouting for southern corn rootworm requires excavation of trenches 4–5 inches (10–12 centimeters) deep and 9 inches (23 centimeters) long at the base of peanut plants. Five such "digs" are needed for each 4–5 acres of a field. Soil must be dug out from under plant parts to expose some of the roots, pegs, and pods for inspection. When insect feeding damage is found, the site is considered infested, regardless of the number of larvae present.

Effective insecticides are available for management of southern corn rootworms. These insecticides are normally applied over the row in a 12- to 16-inch (30- to 41-centimeter) band. The developing pods are protected after the insecticide is incorporated into the soil by rainfall or irrigation. Typically, insecticides are applied during early pod development, but recent research has shown that at least some labeled insecticides provide residual control when applied during flowering. Additionally, most of the insecticides are effective against other pests, including the lesser cornstalk borer, cutworm, and potato leafhopper. Although this broad spectrum of activity has advantages, it may be detrimental to beneficial organisms and may enhance development of corn earworms and spider mites. Research is needed to define nontarget effects of preventive-use insecticides for control of southern corn rootworm.

The cultivar NC 6 was developed for its resistance to southern corn rootworm (Box 8.3). Production of NC 6 in the Virginia-Carolina production region can minimize losses from the pest. When NC 6 is produced in fields with high organic matter and clay content, the insecticide dosage may be reduced to one-fourth the normal amount. In moderate- to low-risk fields, NC 6 virtually eliminates the need for insecticides. No cultural controls are available for southern corn rootworm; however, poorly drained and wet fields should be avoided or treated with insecticide.

Lesser Cornstalk Borer

Economically damaging populations of the lesser cornstalk borer (*Elasmopalpus lignosellus*) can occur at any time during the growing season. The caterpillars inhabit the upper soil and feed on plant structures at or just below the soil surface. The larvae live in silken tubes attached to the peanut plant just below the soil surface (Plate 66). Larvae are small (up to 5/8 inch long) and have alternating aqua blue and purple bands on each segment. The larvae bore tunnels into peanut stems, impairing water and nutrient uptake. Early season infestations result in reduced plant stands and weakened plants, and plants injured by borers wilt readily. During flowering, the larvae attack the main stems of plants, thereby leading to direct pod yield loss and indirect losses by predisposing the plant to disease. The greatest economic losses result from larval feeding on pegs and pods late in the season, again from direct yield losses and indirectly from the introduction of soil microorganisms. Aflatoxin contamination can be a very serious result of pod injury.

Economically damaging populations frequently occur in peanut fields during periods of hot, dry weather. High soil temperatures, typical during drought, favor population development and survival. All commercial cultivars are highly susceptible to lesser cornstalk borer damage. However, planting early in the spring and using early maturing cultivars will reduce the probability of economic damage. During early vegetative crop growth stages, damage can be reduced by keeping the field free of weeds and grass for several weeks prior to planting. Irrigation can be used to reduce economic losses from lesser cornstalk borer.

The primary tactic for suppressing larval populations is control with insecticides. The available efficacious insecticides are usually applied as granular formulations. This strategy allows the chemical to be applied in a 12- to 16-inch (30- to 41-centimeter) band over the row and then to fall to the soil surface. Rainfall or irrigation is needed to incorporate the insecticide into the appropriate upper layer of soil.

Irrigation can be used to reduce economic losses from the lesser cornstalk borer.

The most commonly used soil insecticides for lesser cornstalk borer control are chlorpyrifos and fonofos. These chemicals are readily incorporated by water into the soil habitat, where they remain until degraded. This is desirable because the chemicals remain in the habitat where control is effective and do not pollute groundwater or runoff water. Both products have a fairly long period of residual activity. Control with chlorpyrifos can be expected to last about 28–60 days or longer. The residual efficacy of fonofos is shorter. Residual activity of both insecticides is reduced in hot, dry weather.

Treatment for the lesser cornstalk borer is usually recommended when 10–20% of the plants are infested.

Scouting to estimate the population density and treating only when lesser cornstalk borer populations reach economic threshold levels are not always reliable. Therefore, the curative approach may not be successful in rain-fed peanut fields. Scouting for this pest is laborious and technically difficult. Trenches, similar to those dug for southern corn rootworm, must be carefully excavated because the silken tubes are delicate and often have soil stuck to them. Lesser cornstalk borer populations sometimes develop rapidly. The curative approach, however, can be employed with moderate success when peanut is produced with irrigation. Economic thresholds vary somewhat among states and are usually based on the percentage of sample sites (length of row or number of plants) with larvae or recent damage. The recommended threshold for treatment depends on crop growth stage. However, treatment for the lesser cornstalk borer is usually recommended when 10–20% of the plants are infested.

The pest is most reliably controlled when insecticides are applied before populations reach outbreak densities. Because rainfall or irrigation is needed before the insecticides become effective, applications made after populations reach outbreak densities frequently do not prevent unacceptable economic damage. In addition to controlling lesser cornstalk borer and other soil insect pests, most of the labeled granular insecticides reduce losses from soil diseases such as southern stem rot caused by *Sclerotium rolfsii*. Because of this added benefit, an application at the early seed development stage frequently results in a positive net economic return, even when populations of individual soil insect pests do not reach economic threshold densities. For this reason, there has been a growing trend toward the preventive application of a soil insecticide during early seed development. This practice is increasing, regardless of whether the crop is produced with irrigation or without.

The compatibility of preventive soil insecticide application with IPM programs of peanut has not been adequately assessed. Research has revealed that an application of chlorpyrifos at the early seed crop growth stage reduces populations of spiders, red imported fire ants, and earwigs, but populations of damsel bugs, bigeyed bugs, and ground beetles are not greatly affected. As a result, injury from lepidopterous pests sometimes increases because of chlorpyrifos application, but damage is not usually at economic levels. Effects of chlorpyrifos on populations of other foliage-inhabiting pests have not been determined. Preventive use of soil insecticides may be compatible with IPM programs in some situations. However, more research on use the of soil-applied insecticides is needed.

Wireworms and Whitefringed Beetles

Both wireworms (*Conoderus* spp.) and whitefringed beetles (*Graphognathus* spp.) occur sporadically but are sometimes devastating in individual peanut fields. In general, widespread economic problems are rare in all peanut-producing areas, but there is limited information available concerning their economic impact on peanut production. Economic losses from wireworms result from fruit-feeding activities, but other plant structures are also injured. The immature stage of whitefringed beetles injures young peanut plants by feeding on underground plant structures; the injury sometimes results in the cutting of the taproot. Injury usually results in stunting of the plant, but sometimes the plant dies. For both pests, little information exists concerning the effects of injury at any crop growth stage on peanut yield or quality.

Wireworms are the larval stage of click beetles, which always occur in the soil. Immature insects are present in most peanut fields, and their presence does not indicate an economically important infestation. Their life cycle ranges from 1 to 3 years. Economic problems usually occur when peanut follows the preferred hosts: tobacco, pasture, or sod. Cultural control is effective for wireworms. Waiting at least 2 years after tobacco, sod, or pasture production will reduce the likelihood of economically damaging wireworm problems. The yield benefit from planting peanut after pasture may exceed the detrimental effects of wireworm damage.

In a field in which a high risk is suspected, the soil may be sieved to determine the wireworm population prior to planting. Under high-risk situations, insecticide treatment is sometimes recommended. For example, Georgia's recommendations include incorporation of a soil insecticide when one or more wireworm larvae per square foot of soil surface are found.

The larvae of whitefringed beetles are small (about 0.5 inch long) and legless. The head is inconspicuous and can be detected only by the presence of dark-colored jaws. The pest overwinters as eggs or larvae in the soil. Adults emerge from May through September and feed on foliage of peanut and other plants. This injury is never economically important. The presence of large numbers of beetles in a field may indicate larval problems the following year. Unfortunately, there are no known effective controls for the larval stage of this pest.

Selected References

Funderburk, J. E., and Higley, L. G. 1993. Management of arthropod pests. Pages 199-228 in: Sustainable Agriculture Systems. J. L. Hatfield and D. L. Karlen, eds. Lewis/CRC Press, Boca Raton, FL.

Funderburk, J. E., Higley, L. G., and Buntin, G. D. 1993. Concepts and directions in arthropod pest management. Adv. Agron. 51:125-172.

Hutchins, S. H., and Funderburk, J. E. 1991. Injury guilds: A practical approach to managing pest losses to soybean. Agric. Zool. Rev. 4:1-22.

Pedigo, L. P., and Buntin, G. D., eds. 1993. Handbook of Sampling Methods for Arthropods in Agriculture. CRC Press, Boca Raton, FL.

Pedigo, L. P., Hutchins, S. H., and Higley, L. G. 1986. Economic injury levels in theory and practice. Annu. Rev. Entomol. 31:341-368.

Smith, J. W., Jr., and Barfield, C. S. 1982. Management of preharvest insects. Pages 250-325 in: Peanut Science and Technology. H. E. Pattee and C. T. Young, eds. American Peanut Research and Education Society, Yoakum, TX.

CHAPTER NINE

J. L. Sherwood
Department of Plant Pathology
Oklahoma State University, Stillwater

H. A. Melouk
U. S. Department of Agriculture
Agricultural Research Service
Stillwater, Oklahoma

Viral Diseases and Their Management

Viruses are among the simplest in composition of the pathogens that infect peanut. They consist primarily of nucleic acid and protein. The nucleic acid carries the information to cause disease. The protein coats the nucleic acid to protect it and is involved in the uptake and transmission of the virus by vectors. Vectors are the insects, nematodes, and fungi that carry the virus from plant to plant. Most viruses are transmitted by one or a few species of a specific vector. Although viruses are simple in composition, they are probably the most difficult pathogens to manage and control. There are no curative agents that can be applied to plants to control virus diseases, so management practices must be directed toward preventing the crop from becoming infected or toward keeping initial disease from spreading. In developing management strategies for control of plant viruses, an understanding of the relationship of the virus with the crop, the environment, and the vector is necessary (Boxes 9.1 and 9.2).

Box 9.1

Components of Virus Diseases and Transmission of Viruses

For disease to become established and spread, there must be an initial source of virus. This may be within the crop or outside the crop. Potential sources of virus from inside the crop are virus-infected seed, infected perennial weeds growing in the field, or a viruliferous vector, such as an insect, that has become established in the field. What controls the transission of viruses through seed is not known. For those viruses that are transmitted by seed, the amount of seed that becomes infected on a virus-infected plant can vary from none to a high percentage. The cultivar, the time of infection, seed storage conditions after harvest, and the virus strain affect the percentage of infected seed. Virus transmission through seeds results in the establishment of a disease in a field early in the growing season. If a vector is present to transmit the virus from this primary source of inoculum, the disease may spread rapidly. Sources outside the crop include infected plants from which vectors acquire the virus and bring it into the field. These vectors may have moved long distances.

A vector is required for the spread of virus from plant to plant during the growing season. The most important vectors of viruses of peanut in North America are aphids and thrips (Chapter 8). The relationship between plant viruses and their insect vectors is quite complex. Most viruses of peanut transmitted by aphids are transmitted in a nonpersistent fashion, while viruses that are transmitted by thrips are propagative in the insect.

The environment plays a significant role in the interactions of the components of the disease. For control to be achieved, one of the disease components has to be modified or eliminated. This could include the elimination of the virus source, use of a resistant cultivar, control of the vector, or modification of the environment. All elements of the disease are important and interrelated.

Viruses that Affect Peanut in North America

Peanut growers are fortunate in that the number of viruses dramatically affecting peanut production in North America is limited. The viruses associated with peanut have been characterized and are members of the plant virus families Comoviridae, Bromoviridae, Potyviridae, or Bunyaviridae. A virus generally receives its name on the basis of the host on which it was initially found and a symptom it causes. However, this is not always the case. Plant viruses can be identified by serological assays such as the enzyme-linked immunosorbent assay (ELISA) (Box 9.3). This service is generally available through county extension offices or state land grant universities.

Peanut Mottle Virus

Peanut mottle virus (PMV) was first recorded in 1965 in Georgia and has since been found worldwide in all major peanut-producing areas. PMV is a member of the Potyviridae. Symptoms can vary with cultivar, time of infection, and environment and are generally mild, which may explain why such a widely distributed pathogen was not detected until 1965. The most common symptom is a mild mottle or mosaic on the youngest leaves of infected plants (Plate 67). Margins of leaflets may curl up, and depressions in the interveinal tissue may become prominent. Plants are generally only mildly stunted, if at all. As plants mature, the symptoms usually wane. Pods from infected plants may be reduced in size and have irregular gray to brown patches. The seed coat of affected seed also may be discolored.

Other leguminous hosts of PMV are soybean, lupine, bean, pea, clover, and cowpea. Although PMV can be isolated from some weed species, they are not thought to play significant roles in the epidemiology of peanut mottle.

Disease Initiation, Development, and Spread. The virus is readily transmissible by mechanical means, which makes detection with diagnostic hosts such as the bean cultivar Topcrop possible. PMV is transmitted in a nonpersistent manner by aphid species including the cowpea, cotton, green peach, bird cherry-oat, and corn leaf aphids.

PMV is transmitted from the maternal plant to the seed. Infected seed is thought to be the primary source of virus for the beginning of the disease cycle. Although only about 1% of the seed from an infected plant carries the virus, plants from infected seed can serve as the initial source from which aphids can very efficiently spread the virus.

Winter forage legumes are also hosts for PMV and may serve as sources of the virus. However, there is not an abundant acreage of these crops in many peanut-growing areas in the United States.

Disease Losses and Control. As with the severity of symptoms, the losses caused by PMV will vary depending on the cultivar, time of infection, and strain of the virus. PMV has been estimated to cause crop losses of about 5%. Although several plant introductions with resistance to PMV have been identified, there are no peanut cultivars with resistance to the virus. The use of PMV-free seed is the most feasible approach for control, but because of the lack of certification programs, the best chance of obtaining virus-free seed is to purchase the highest quality seed available. The use of virus-free seed prevents the disease from becoming initially established in the field. Additionally, where winter forage legumes are grown, the peanut fields should be as far as possible from these fields. The virus can be detected with serological assays.

Peanut Stripe Virus

Peanut stripe virus (PStV) was first detected in the United States in Georgia in 1982. It had apparently been introduced through seed received from the People's Republic of China. Peanut stripe has not become a significant problem in peanut production in the United States, but the virus can be an important pathogen in other areas of the world. PStV is a member of the Potyviridae.

Box 9.2

Some Types of Transmission of Viruses by Insects

Nonpersisent

The insect acquires the virus from the plant and is able to transmit it immediately after acquisition. The insect will be able to transmit the virus only for a short time unless the virus is acquired again.

Persistent

The insect acquires the virus from the plant, and there is a period (hours to days) before it is able to transmit the virus (latent period). The insect will be able to transmit the virus for many days up to the life of the insect. This type is also referred to as circulative transmission.

Propagative

This type is similar to persistent transmission except that the virus replicates in the insect.

Box 9.3

Summary of the Serological Assay for Plant Viruses

When the immune system of an animal encounters a foreign protein, one response is the production of antibodies. Plant viruses, of which a large component is protein, can be administered to an animal, generally a laboratory rabbit, which will produce the immune response and the antibodies to the plant virus. The serum fraction containing the antibody to the virus can be isolated from the blood of the animal. This antiserum is then used in serological tests with sap from plants to test for the presence of the plant virus. The antibodies to the plant virus in the antiserum will specifically bind to the virus if it is present in the plant sap. This binding can be detected in a variety of ways. The advantages of using a serological assay are that results can be obtained in a day or two and a large number of samples can be run in a short period of time. Most virus diseases are diagnosed by this method.

Symptoms in leaves after initial infection consist of a striping or banding along the lateral veins of leaflets (Plate 68). As plants mature, the striping pattern fades and a pattern similar to the outline of an oak leaf may become evident. Although the virus has been found in samples from Florida, Georgia, North Carolina, New Mexico, Oklahoma, Texas, and Virginia, the distribution has been primarily limited to germ plasm in research plots. Efforts to prevent wide dissemination of the virus in the United States seem to have been effective. In addition to China, India and Southeast Asia have reported PStV.

> ## Peanut stripe has not become a significant problem in peanut production in the United States.

Disease Initiation, Development, and Spread. As with PMV, PStV is mechanically, aphid, and seed transmitted. The cowpea and green peach aphids have been shown to transmit PStV in a nonpersistent fashion. An early report on seed transmission of PStV in the greenhouse indicated that the virus may be transmitted to up to 37% of seed. A subsequent study in the United States found transmission of PStV to seed of infected plants to be approximately 5% or less. However, PStV was transmitted in up to 50% of the seed in some genotypes in China.

Disease Losses and Control. Under controlled field conditions in the United States, PStV seems to have little effect on growth, yield, or grade of Florunner peanut. Resistance to PStV has been identified in some peanut-related germ plasm, but the germ plasm lines are not readily crossed with cultivars. PStV does not appear to have become a problem in the United States but is an important virus in other areas where peanut is grown. In China, yield reductions of up to 23% have been reported. No resistant cultivars have been reported. Virus-free seed is recommended to eliminate a possible source of the virus. The virus can be detected with serological assays.

Peanut Stunt Virus

Peanut stunt virus (PSV) was first observed in peanut in the United States in Virginia and North Carolina in 1964. Although of significant importance in peanut during 1964 and 1965, PSV is not now an important pathogen of peanut in North America. The virus is found in white clover and other legumes primarily in the southeastern United States but also has been found in the Midwest in white clover. PSV has also been reported in France, Hungary, Japan, Morocco, Poland, Spain, and Sudan. PSV is a member of the Bromoviridae.

> ## PSV has not been reported to be a significant problem in peanut since the 1960s.

The most common symptom of infection is severe stunting, which may affect the entire plant or only a portion, such as a single branch. The earlier the virus infection, the greater the severity of the symptoms of the disease. Leaves of infected plants may be curled upward and malformed. Leaves may also be chlorotic or mottled or have a dark green banding pattern. Pod size and number are both reduced. Pods also may be malformed and split. Seed from infected plants have reduced viability and produce seedlings that lack vigor.

PSV has a wide host range that includes numerous species in the legume family and approximately 58 other genera of plants. Important economic hosts include peanut, bean, cowpea, and white clover.

Disease Initiation, Development, and Spread. The virus can be transmitted mechanically, by aphids, and in seed. Seed transmission is not thought to be of significance in disease spread because the percentage of seed transmission is less than 0.3% and the viability of plants from infected seed is low. Aphids, however, play an important role in the transmission of the virus. The green peach, cowpea, and green citrus aphids transmit PSV in a nonpersistent manner. Aphids may acquire the virus from infected hosts such as white clover or weeds such as crownvetch and transmit it to nearby peanut fields.

Disease Losses and Control. PSV has not been reported to be a significant problem in peanut since the 1960s. Information from that time indicates that losses of up to 75% could occur and the quality of the peanuts produced was low. Control or management of this virus disease may be aided by planting peanut away from fields of white clover and by the use of peanut seed free of PSV. Cultivars with resistance to PSV have not been developed, although some germ plasm lines with resistance have been identified. The virus can be detected by serological assays.

Tomato Spotted Wilt Virus

Tomato spotted wilt virus (TSWV) was first observed in peanut in Brazil in 1941 and has since become an important pathogen in many peanut-growing areas worldwide. In the United States, TSWV is generally found wherever peanut is grown, and in some years in some locations, severe epidemics have developed. In India, bud necrosis disease is caused by a virus similar to TSWV and is probably the most important virus disease of peanut. TSWV is a member of the Bunyaviridae.

The symptoms of TSWV in peanut can be quite varied. The symptom most closely associated with the virus in peanut is the development of chlorotic ring spots of various sizes and shapes (Plate 69). Stunting also occurs and can be severe if plants are infected early in the growing season (Plate 70). In Georgia, symptoms produced by some strains of PMV and TSWV in peanut are similar. Terminal bud necrosis is commonly observed in infected peanut in India. Pod size and number are reduced, especially if plants are infected early. Seed produced on infected plants may be reduced in size and malformed and have discolored (red) seed coats (Plate 71). Seed transmission of the virus has not been reported.

Disease Initiation, Development, and Spread. TSWV has one of the widest host ranges of all plant viruses. A complete list has yet to be developed, but approximately 200 species of host plants have been reported. The virus can be mechanically transmitted to petunia for diagnostic purposes, but thrips are the primary vectors of TSWV. Those that transmit TSWV include the common blossom, tobacco, western flower, chilli, and onion thrips. The virus is acquired by larvae and transmitted when these larvae become adults. TSWV is propagative in thrips when acquired by larvae. Adults that did not acquire the virus as larvae can take up the virus when they feed on TSWV-infected plants, but they do not transmit the virus to other plants when they feed.

Generally, the earlier the peanut crop is exposed to TSWV,

the more severe the effects of the disease. Several weed species are hosts of TSWV, and although virus-infected weeds may serve as a source of the virus, their importance in the disease cycle is not certain. Since thrips are generally an important insect problem in peanut, the potential exists for TSWV to be an important disease.

Generally, the earlier the peanut crop is exposed to TSWV, the more severe the effects of the disease.

Disease Losses and Control. Losses to TSWV can be severe and are often sporadic. During the mid-1980s, losses in some fields in Texas were 100%, but within 2 years, the disease had significantly declined. Because of the wide host range of the virus, the potential exists for severe epidemics and significant losses. TSWV can be detected by serological assays.

Cultivars with complete resistance to TSWV are not available. Southern Runner and Georgia Browne have been found to have some resistance to TSWV. Insecticides will aid in the control of thrips, but there is no definitive work to indicate that control of thrips with insecticides will control TSWV. The control of weeds may or may not help in disease control.

Other Viruses

Viruses of Minor Importance in North America

Bean yellow mosaic virus (BYMV), a member of the Potyviridae, was identified in peanut in Georgia in 1986. Chlorotic spots and rings occur in leaves of infected plants, but the symptoms fade as the plant matures. The virus can be transmitted by the cowpea aphid. In a survey conducted in 1986, the virus was found in only one location. BYMV was not found to be seed transmitted in peanut and caused no apparent yield loss in a greenhouse test.

Cowpea chlorotic mottle virus (CCMV), a member of the Bromoviridae, has been found on several occasions in peanut in Georgia. The virus does not cause visible symptoms in infected plants. Although CCMV does not cause disease by itself, it may when in mixed infections with other viruses.

Viruses of Major Importance Outside North America

With the continual movement of plant material and soil between nations, deliberate or otherwise, there is always the possibility that a new virus that affects peanut will enter North America. There are two viral diseases of significance in peanut-growing areas outside North America: groundnut rosette and peanut clump.

Groundnut rosette was first reported in 1907 and may be limited to Africa. It is caused by a complex consisting of two viruses, groundnut rosette virus and groundnut rosette assistor virus, and a small piece of virus-associated nucleic acid called a satellite. The complex is transmitted by the cowpea aphid. Infected plants have a general chlorosis and faint mottling of young leaflets. Newly formed leaflets are small, chlorotic, curled, and distorted. Older leaves may appear yellow with green patches. Cultivars that express resistance to this complex have been reported.

Peanut clump virus (PCV) has been reported in several locations in India and West Africa. In both geographic areas, it is transmitted by a fungus that is a common inhabitant of the soil. Once an area becomes infested with the viruliferous vector, it generally remains a source for the virus. The symptoms seen in affected areas are typical of many soilborne diseases. The disease occurs in localized areas in a field, and the affected areas enlarge progressively during the following years. Newly formed leaves on infected plants may appear mottled and have chlorotic ring spots. These symptoms fade, and the leaves become dark green. Infected plants are severely stunted. Although flowers and pegs are produced, the number and size of pods produced on infected plants are severely limited. The root system of the infected plant is also adversely affected. Some tolerance to PCV has been reported in breeding lines in India.

Strategies for Management of Virus Diseases

Although viruses of peanut can cause losses, those affecting peanut production in North America and their severity are limited (Table 9.1). Before controls are implemented, one must be sure the disease is caused by a virus. The symptoms produced by virus infection can be similar to those caused by nutrient imbalance, herbicide injury, or insect damage. Samples of affected peanut should be tested for virus infection, a service usually available through extension offices or land grant universities.

The symptoms produced by virus infection can be similar to those caused by nutrient imbalance, herbicide injury, or insect damage.

The strategies for management of virus diseases are similar to those for other pathogens of peanut. There are several approaches that can be used, and it is best if more than one is employed. The use of cultivars with resistance is the easiest, but the availability of these cultivars is limited or nonexistent. Efforts are underway in most peanut-growing areas to develop

Table 9.1. Viruses that have significantly affected peanut in North America

Virus	Main modes of transmission	Frequency	Severity
Peanut mottle virus	Aphids, seeds	Common	None to moderate
Peanut stripe virus	Aphids, seeds	Rare	Moderate
Peanut stunt virus	Aphids	Rare	Moderate to severe
Tomato spotted wilt virus	Thrips	Common	Moderate to severe

cultivars resistant to endemic viruses of recurring importance, and progress has been made in developing cultivars with resistance to TSWV. Modification of planting date or planting area, if possible, to avoid viruliferous insects is another approach to facilitate control. Other potential control methods are intercropping with plants that are not virus hosts and the use of wind breaks to reduce vector movement. The use of virus-free seed to control seed-transmitted diseases is a good approach, although certification programs for production of virus-free seed have not been developed in North America. It is best to plant the highest quality seed available. Control of weeds and insects may aid in the control of virus diseases but should not be the primary approach.

The utility of each of these tactics will vary with the virus, the geographical area, and the cultivars. Extension offices can provide the most up-to-date information on diseases and control strategies in specific regions.

Selected References

Bock, K. R., and Kuhn, C. W. 1975. Peanut mottle virus. Descriptions of Plant Viruses, no. 141. Commonwealth Mycological Institute and Association of Applied Biologists, Kew, England.

Demski, J. W., and Lovell, G. R. 1985. Peanut stripe virus and the distribution of peanut seed. Plant Dis. 69:734-738.

Francki, R. I. B., and Hatta, T. 1981. Tomato spotted wilt virus. Pages 491-512 in: Handbook of Plant Virus Infections and Comparative Diagnosis. E. Kurstak, ed. Elsevier-North Holland Biomedical Press, New York.

Ie, T. S. 1970. Tomato spotted wilt virus. Descriptions of Plant Viruses, no. 39. Commonwealth Mycological Institute and Association of Applied Biologists, Kew, England.

Kucharek, T., Brown, L., Johnson, F., and Funderburk, J. 1990. Tomato spotted wilt virus of agronomic, vegetable, and ornamental crops. Fla. Coop. Ext. Serv. Circ. 914.

Lynch, R. E., Demski, J. W., Branch, W. D., Holbrook, C. C., and Morgan, L. W. 1988. Influence of peanut stripe virus on growth, yield, and quality of Florunner peanut. Peanut Sci. 15:47-52.

Matthews, R. E. F. 1981. Plant Virology. Academic Press, New York.

Mink, G. I. 1972. Peanut stunt virus. Descriptions of Plant Viruses, no. 92. Commonwealth Mycological Institute and Association of Applied Biologists, Kew, England.

Porter, D. M., Smith, D. H., and Rodríguez-Kábana, R. 1982. Peanut plant disease. Pages 326-410 in: Peanut Science and Technology. H. E. Pattee and C. T. Young, eds. American Peanut Research and Education Society, Yoakum, TX.

Porter, D. M., Smith, D. H., and Rodríguez-Kábana, R., eds. 1984. Virus diseases. Pages 45-52 in: Compendium of Peanut Diseases. American Phytopathological Society, St. Paul, MN.

F. W. Nutter, Jr.
Department of Plant Pathology
Iowa State University, Ames

F. M. Shokes
North Florida Research and Education Center
University of Florida, Quincy

CHAPTER TEN

Management of Foliar Diseases Caused by Fungi

Foliar diseases of peanut remain among the most important yield-limiting factors in peanut production. They reduce the amount of healthy leaf area available for photosynthesis, resulting in a decreased ability to produce and convert the products of photosynthesis into high-quality peanuts. Plant vigor may be reduced if young seedlings become infected. There is also evidence that foliar diseases, when severe, can reduce the percentage of sound, mature kernels.

The foliar fungal diseases of peanut that occur in the United States include early leaf spot, late leaf spot, rust, web blotch, Phomopsis leaf spot, Alternaria leaf spot, leaf scorch, and pepper spot. The most troublesome are early leaf spot and late leaf spot, and a large portion of the pesticides used on peanut are expended for management of these two diseases. Therefore, a major portion of this chapter is devoted to the management of early and late leaf spot diseases.

Of necessity, when plant diseases, the infection process, and disease cycles are discussed, a few terms may be used that are unfamiliar to the average reader. Therefore, a short list of definitions is given in Box 10.1.

Early and Late Leaf Spot Diseases

Early and late leaf spots are two of the most important foliar diseases caused by fungi wherever peanut is grown throughout the world. When fungicidal control is not used with susceptible cultivars, yield losses from these diseases may approach 70%. Although both leaf spot diseases can be found throughout the peanut-growing regions of the United States, early leaf spot, caused by *Cercospora arachidicola* S. Hori, is more prevalent in Virginia, North Carolina, Oklahoma, and New Mexico (Fig. 10.1). Late leaf spot, caused by *Cercosporidium personatum* (Berk. & M. A. Curtis) Deighton, was predominant in Georgia, Florida, and Alabama throughout the 1980s. In recent years, however, early leaf spot has become more prevalent in Georgia and Alabama, and now both diseases occur each year. Both leaf spots are equally prevalent in the peanut-growing areas of Texas and South Carolina. Late leaf spot predominates in southern peanut-growing regions of Texas, and early leaf spot is prevalent in the northern peanut-growing areas of that state. Although there may be a prevalence of one leaf spot pathogen in a given area, the predominance of one over the other can shift in any given year or over several years in response to changes in climate. For example, during 1992 and 1993, early leaf spot remained predominant throughout the season in many parts of Georgia and Alabama.

Box 10.1

Definitions

conidia   asexual spores
conidiophore   a specialized hypha on which conidia are produced
defoliation   shedding of leaves
haustoria   simple or branched fungal structures that invaginate cells and absorb nutrients from them
incubation period   the time from infection to appearance of symptoms
inoculum   that portion of a pathogen that may be transferred to a plant to cause an infection, e.g., spores
intercellular   between the cells
mycelium   the filamentous vegetative growth of a fungus; hypha
pathogen   a causal agent of disease in a plant
stoma   a minute opening in the plant epidermis
stroma   a mass of fungal tissue in which spores are produced

---

Symptoms

The symptoms of each leaf spot disease are unique. The diseases are probably most destructive when they occur on peanut leaves, but all of the aboveground plant parts are subject to infection. It is not uncommon to find lesions on leaf petioles (stems), pegs, central stems, and lateral branches late in the season. Lesions are dead areas of tissue that develop on leaves (Plate 72) after infection by either of the leaf spot fungi. During the initial stages of development, lesions appear as tiny, pinpoint, yellowish specks.

Early leaf spot develops into irregularly shaped to circular spots that are about the size of a pencil eraser. The color varies from dark brown to almost black on the upper surfaces of leaflets (Plate 72) and is generally tan to reddish tan on the lower leaflet surfaces of susceptible cultivars (Plate 73). A yellow halo often surrounds the dark spot and is more pronounced on the upper surface. This yellowing is caused by a toxin produced by the early leaf spot fungus. The toxin kills cells in advance of the fungal growth, resulting in the yellow halo. Spores (conidia) are produced on stalks (conidiophores) mainly on the upper surfaces of lesions, but a few may be found on the lower surfaces of older spots. The spores grow out of a fungal stroma that looks like a pinhead-sized black spot on the lesion. Spores are colorless to slightly olive and look almost white to the naked eye or when viewed with a hand lens.

Lesions of late leaf spot are also circular but are not as irregular as those of early leaf spot. Mature spots tend to be slightly smaller than those of early leaf spot (about 70% the size of a pencil eraser). Yellow halos may or may not occur with late leaf spots. When halos do occur, they are a pale yellowish green on the lower leaflet surface but may be bright yellow on the upper surface. Unlike the early leaf spot fungus,

the late leaf spot fungus produces specialized feeding structures called haustoria (Fig. 10.2). These structures allow the leaf spot fungus to parasitize the plant cells without immediately killing them.

Late leaf spots are dark brown to almost black on the upper leaf surface. In the initial stages of development, it is difficult

---

Box 10.2

Partial Resistance to Leaf Spot Diseases

None of the commercial peanut cultivars has complete resistance to leaf spot diseases. Therefore, all of them will develop some level of leaf spot, but the disease in those with partial resistance will typically be less severe. Such cultivars often do not have good resistance to infection by the pathogen, and plants may develop many lesions if high levels of inoculum are present. That is, this resistance is not usually expressed in peanut as fewer lesions but in reduced size of lesions and reduced numbers of spores produced in those lesions by the fungus. In addition, the pathogen usually has a longer latent period (the time from infection to spore production by the fungus on a lesion) on a partially resistant cultivar. These characteristics are additive in decreasing the number of disease cycles (Fig. 10.2) that occur during a season and will collectively lessen the severity of disease within the field. This type of resistance is often called "rate-reducing" resistance because it decreases the rate of development of the disease epidemic.

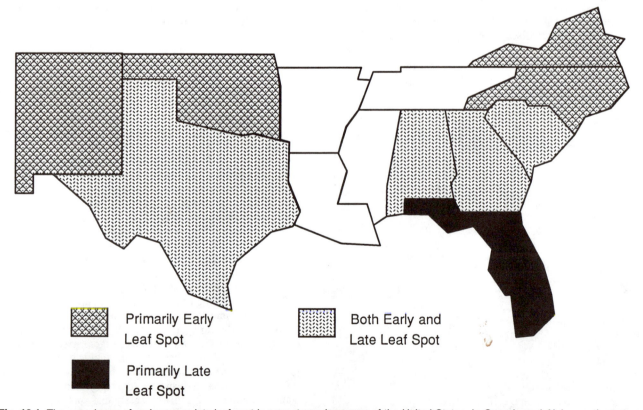

Primarily Early Leaf Spot

Both Early and Late Leaf Spot

Primarily Late Leaf Spot

Fig. 10.1. The prevalence of early versus late leaf spot in peanut-growing areas of the United States. In Georgia and Alabama, there was a predominance of late leaf spot during the 1980s, but there has been a recent shift in the prevalence of early leaf spot.

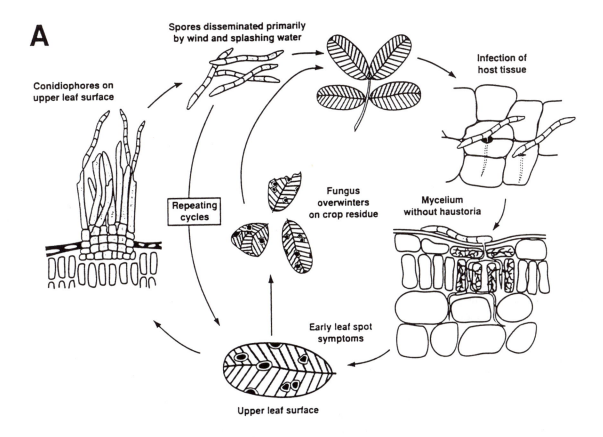

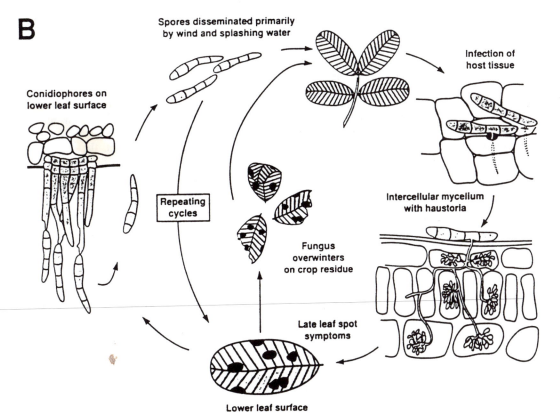

Fig. 10.2. Disease cycles of **A,** early leaf spot, caused by *Cercospora arachidicola* and **B,** late leaf spot, caused by *Cercosporidium personatum*.

to differentiate between early and late leaf spots. As late leaf spot lesions mature, they become dark brown to black on the underside of the leaflet surface as well (Plate 74). Sporulation occurs mainly on the lower leaflet surface, although some spores may also be produced on the upper surface of older lesions. In contrast to the scanty numbers and whitish appearance of early leaf spot spores, those of late leaf spot are a dark olive brown and occur in thick tufts. These olive brown tufts often occur in concentric circles and are readily seen with the naked eye (Plate 75), appearing as small bumps on the lesions.

The color of early and late leaf spot lesions is not a reliable characteristic for distinguishing between them on partially resistant cultivars such as Southern Runner (Box 10.2). Because there is so much variation in lesion characteristics on resistant cultivars, incubation in the laboratory coupled with microscopic examination may be required to determine which disease is present. If the leaf spots are sporulating, a hand lens may be all that is necessary for positive identification by an experienced observer.

Disease Cycles

Both leaf spot pathogens can reproduce and spread disease very rapidly. This causes severe epidemics, which occur quickly and cause significant defoliation (leaf shedding) and yield loss (Plate 76). It is believed that both pathogens survive intercrop periods in crop residue (Fig. 10.2). As soon as the crop emerges from the soil, spores disseminated by wind and rain splash are deposited on peanut leaf surfaces. When weather is favorable for infection, spores will germinate and penetrate host tissue primarily through stomata (leaf pores). Both leaf spot pathogens grow within the leaf tissue, and the late leaf spot fungus produces the fingerlike feeding structures called haustoria (Fig. 10.2). Early leaf spot lesions and halos can be seen in 6–8 days but are more commonly detected 11–17 days after infection has occurred. Late leaf spot lesions can be detected 10–14 days after infection.

Typically, early leaf spot epidemics are favored by temperatures of approximately 61–77°F (16–25°C) and long periods of high relative humidity over several days. Conditions favorable for late leaf spot epidemics include wet periods with temperatures of 68–79°F (20–26°C). Prolonged periods of leaf wetness or several shorter periods of leaf wetness (10 hours or longer) may be equally favorable for late leaf spot development. During warm, wet weather, new spores will be produced on lesion surfaces. In spite of having a longer incubation period (the time from infection to appearance of lesions), the late leaf spot fungus generally produces more spores per lesion than the early leaf spot fungus. This is probably the main reason that late leaf spot causes more severe damage over a shorter period of time. Repeated cycles of disease occur during the cropping season. The number of disease cycles depends upon weather, the cultivar grown, and the effectiveness of control measures.

Management Versus Control of Leaf Spot Diseases

Over the years, peanut farming systems have changed in response to the availability of new technologies. The selection of cultivars, fungicides, and tillage systems can all have an impact on disease development. Each management decision made by the grower may hinder or encourage the development of foliar disease epidemics. For this reason, an integrated management system is required. A good example of this is a decision to use chlorothalonil exclusively on a calendar spray schedule (e.g., every 14 days) to control early leaf spot. Exclu-

sive use of chlorothalonil on a calendar schedule may enhance Sclerotinia blight in areas where this disease is a problem. Similarly, higher levels of southern stem rot, caused by *Sclerotium rolfsii* and often called white mold, have occurred in fields of Florunner peanut treated with the fungicide benomyl.

Irrigation is another management practice that may influence disease occurrence and severity. The use of irrigation may be a major factor in maintaining yields during dry years. However, increased use of water can also influence diseases such as early and late leaf spot, resulting in the need for an increased number of fungicide applications to keep the diseases at manageable levels.

Two main strategies are advocated for management of leaf spot diseases. The first is to reduce the level of inoculum during the intercrop period, which will reduce the amount of inoculum available to start an epidemic after the peanut crop emerges. The second strategy is directed toward reducing the rate of increase (number of disease cycles) during the cropping period. There is an important relationship between these two strategies: strategies that reduce inoculum levels will delay the onset of an epidemic. This will become more effective as the rate of spread is reduced during the season. The time delay coupled with the reduction in rate of disease increase can be sufficient to prevent foliar diseases from reaching damaging levels.

Disease-Management Tactics that Reduce Initial Inoculum Levels. Both early and late leaf spot fungi survive between crop periods in plant debris. Disease-management tactics that reduce inoculum in debris have a beneficial effect. Crop rotations of 2 or 3 years out of peanut will result in decreased survival of leaf spot inoculum in the soil. Since both pathogens infect only peanut, crops used in rotation should be selected on the basis of nonsusceptibility to soilborne pathogens that can cause disease on peanut. Appropriate crop rotation can result in a significant reduction in the severity of leaf spot diseases. In field trials where fungicides were not used, defoliation was considerably less when peanut did not follow peanut compared with levels when peanut followed peanut (Fig. 10.3). The time to reach 50% defoliation was delayed from 78 days after planting when peanut followed peanut to 105 days when peanut was in its first year of planting. If the crop had been treated with fungicides to reduce secondary spread of inoculum, the delay would have been

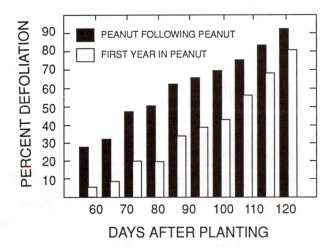

Fig. 10.3. Leaf spot-induced defoliation of peanut plants in relation to crop age for a crop planted following peanut and a crop planted in a field in which peanut had not been grown.

much greater (an additional 20–30 days). Crop rotations of 2 or 3 years would probably provide a significant benefit somewhere between these two extremes. Another tactic that is effective in reducing the amount of inoculum between cropping periods is deep plowing of crop residues. This may also provide some delay in the onset of a leaf spot epidemic but is no substitute for rotation with a nonsusceptible crop.

Perhaps one of the most overlooked tactics is the destruction of volunteer peanut plants or "ground-keepers." The time between successive peanut crops should be as long as possible. It is not unusual for a producer to effectively manage leaf spot during the cropping season only to have great numbers of volunteer plants emerge during the fall or the following spring. Since these volunteer plants are not protected by fungicides, large numbers of lesions may develop, providing inoculum for the next peanut crop. Leaves with sporulating lesions have been collected in Georgia as late as December and January from fall volunteers and as early as mid-April from spring volunteers. This source of inoculum represents an even greater risk in the central Florida growing area. Thus, volunteer plants can be an important source of inoculum for both leaf spot pathogens to bridge the period between peanut crops.

Disease-Management Tactics that Reduce the Rate of Disease Increase. Cultivars with partial resistance (Box 10.2) to early or late leaf spot will reduce the rate of disease increase. However, the level of resistance in current commercial cultivars is not yet high enough to totally eliminate the need for fungicidal protection. Fungicides provide the most effective way to reduce the rate of foliar disease increase on peanut. Until recently, chlorothalonil was the most widely used fungicide for management of foliar diseases on peanut in the United States. It is important to consult with local extension personnel for recommendations when choosing a fungicide because the labels change frequently.

Typically, fungicides are applied either on a calendar spray schedule or according to weather-based spray advisories. When fungicides are applied on a calendar schedule, it is usually recommended that applications be made at 10- to 14-day intervals, depending on the weather. Fungicide application is usually recommended at 14-day intervals or, if rainfall is frequent, at 10-day intervals. This type of program is aimed at preventing disease and maintaining protection throughout the growing season. The selected intervals allow periodic protection of new foliage and take into account the length of the repeating cycles for both leaf spot diseases. When a calendar schedule is used in the Georgia-Florida-Alabama growing region, it is suggested that fungicide applications begin about 30–40 days after planting. In the Virginia-Carolina growing area, the calendar schedule suggests beginning leaf spot fungicide applications between June 25 and July 1. In Texas, it is suggested that the first scheduled fungicide applications be made about 60–65 days after planting, unless late leaf spot is present.

Fungicides are often applied with a tractor-mounted sprayer but may also be applied by aircraft, through a center-pivot irrigation system, or by an underslung spray boom mounted on an irrigation system (Chapter 14). Regardless of the application system used, four factors are important for disease management with fungicides: 1) choosing the right fungicide, 2) applying the correct dosage, 3) getting good coverage, and 4) applying the fungicide at the right time. The proper fungicide must be used to be effective in reducing the spread of the fungal pathogen to prevent an epidemic. Similarly, the recommended dosage must be applied according to the instructions on the label.

Both protectant and systemic fungicides are currently available for foliar disease management on peanut. Protectant fungicides serve as chemical barriers to infection at the sites of application. Some limited "kickback" activity may occur with protectants because germinating fungal spores remain exposed on leaf surfaces for a period of time after the infection process has started (usually 72 hours or less).

> **Four factors are important for disease management with fungicides: 1) choosing the right fungicide, 2) applying the correct dosage, 3) getting good coverage, and 4) applying the fungicide at the right time.**

There are three systemic fungicides currently labeled for use on peanut. Benomyl is a systemic fungicide that, unfortunately, cannot be used in many of the peanut-growing regions of the United States because of the widespread development of pathogen resistance to this fungicide. The two triazole (sterol-inhibiting) fungicides, propiconazole and tebuconazole, are now labeled for use on peanut. Systemic fungicides are absorbed by plant tissues and thus provide protection at locations within the plant other than those at which they were first applied. Tebuconazole has been shown to be effective against both leaf spot diseases, peanut rust, and the soilborne pathogens that cause southern stem rot and Rhizoctonia limb rot. Tebuconazole is typically used for control of southern stem rot in some areas but may be substituted for chlorothalonil treatments. Propiconazole is most effective when used against early leaf spot. It may also be used to manage late leaf spot but works best against this disease when tank-mixed with chlorothalonil.

Special care should be taken to use the systemic, sterol-inhibiting fungicides as recommended on the labels. They will often be recommended as tank mixes with chlorothalonil, or it may be suggested that they be used in blocks of three or four applications followed by a chlorothalonil application. These are very important management strategies aimed at preventing the buildup of fungicide-resistant fungal populations, such as occurred with benomyl.

Growers made an average of 2.8 fungicide applications in 1963 and are presently averaging six to seven applications per season. The current calendar system for scheduling fungicide applications could result in seven to nine (or more) applications each season, depending on the initiation date. The development of weather-based models by researchers and the availability of computer data acquisition systems for collection and interpretation of environmental data may accelerate the use of spray advisories by producers. The widespread use of early leaf spot spray advisories in Virginia and North Carolina has resulted in an average of only 4.5 to five sprays per season. It is anticipated that use of one of the late leaf spot spray advisory systems, under development in the Southeast, may also allow an average of four to five fungicide sprays per season. Use of leaf spot advisories in Florida resulted in one to three fewer fungicide applications for managing late leaf spot over 4 years of testing.

The reason that spray advisories reduce the number of fungicide sprays per season is that weather conditions are not always favorable for leaf spot development. Relative humidity, temperature, and leaf wetness are important factors that affect the success of infection by leaf spot fungi. Hot, dry weather can control leaf spot diseases as effectively as fungicides. Calendar-based spraying may provide good disease control but often results in the application of fungicides to the crop when conditions are unfavorable for disease development. The timing of sprays according to spray advisories provides disease management at a lower cost to the producer, since fungicide is applied only as needed. Leaf spot incidence is sometimes slightly higher in fields sprayed according to advisories. However, these higher disease levels are generally insignificant, and higher pod yields have frequently been obtained when a spray advisory was used in place of a calendar schedule. Increased yields obtained by using the advisory programs can be attributed in part to reduced soil compaction and plant damage by tractors. Reduced field traffic can also result in reduced severity of soilborne diseases such as Sclerotinia blight, stem rot, and Rhizoctonia limb rot.

> ## Relative humidity, temperature, and leaf wetness are important factors that affect the success of infection by leaf spot fungi.

More than 60% of the peanut acreage in the United States is located in three states: Georgia, Alabama, and Florida. By using 1993 statistics, it can be estimated that eliminating just one spray application in this three-state area would save about $5 million in fungicide costs alone. When the cost of making the application (gas, oil, and labor) is included ($7.00 per acre), the savings per application would approach $7 million. For example, a grower with 100 acres who sprayed four times

using a spray advisory instead of seven times using a calendar schedule would save approximately $2,100 (three sprays saved × $7.00 per acre per spray × 100 acres). These savings are significant and would increase the net profit per acre. Spray advisories have the added benefit of reducing the amount of pesticide that is applied to the crop and released into the farm environment.

Defoliation and Time of Digging. Research in Georgia and Florida has shown that there is an adverse interaction, at least with leaf spot-susceptible genotypes, between increasing levels of defoliation and time of digging (Fig. 10.4). When the digging date is delayed, more of the mature, highest quality pods are left in the soil. Defoliation induced by leaf spot greatly weakens the branches and underground stems (pegs), causing pod loss during digging. Therefore, it is imperative that growers scout all fields as the crop approaches maturity. An appropriate method, such as the hull-scrape method, should be applied to determine the optimum digging date (Chapter 5). Any fields heavily infected with leaf spot should be checked every 2 days to determine the level of defoliation. Once defoliation exceeds 35% on susceptible cultivars, the crop should be dug as soon as possible to avoid leaving high-quality pods in the soil.

Use of Cultivars with Partial Resistance to Leaf Spot Diseases. Partial resistance to peanut leaf spot diseases (Box 10.2) is another management tool that is now available in a limited number of cultivars grown in the United States (Chapter 4) and should become available in more cultivars in the future. This type of resistance is usually specific for either early leaf spot (e.g., cultivars NC 6 and NC 7) or late leaf spot (e.g., Southern Runner) and not necessarily for both.

Use of partially resistant cultivars within a growing region can decrease the number of fungicide applications required to maintain adequate disease control within a season. For example, four applications of chlorothalonil will maintain adequate control of leaf spot during most years in Southern Runner, a cultivar with partial resistance to late leaf spot. A calendar-based system for this cultivar recommends an initial fungicide application about 60 days after planting and three additional applications at 21-day intervals. By integrating partial resistance with fungicide control, fewer sprays are required to sustain adequate disease management.

Early and Late Leaf Spot Information Delivery Systems. Computerized disease-management delivery systems for early leaf spot were first developed in Georgia in the mid-1960s but have found their greatest success in Virginia and North Carolina where the early leaf spot disease predominates. More recently, a late leaf spot spray advisory model was developed in Georgia that could have widespread use in the lower southeastern states to more efficiently control this disease. This system uses an in-field weather station coupled with a computer (Plate 77). Typical use of this system for spray advisories is described in Box 10.3.

An expert system for growing peanut is under development in Alabama. Part of this system includes a leaf spot control model known as AU-PNUTS, which allows spray decisions to be made on the basis of rain events and requires only the use of an in-field rain gauge. The decision to spray or not to spray is based on the number of previous rain events and the current weather forecasts. The producer must maintain accurate records and keep up with weather forecasts that give the 5-day probability for the occurrence of rain. This system has worked well in tests in the Georgia-Florida-Alabama area.

Mainframe computer-based advisory systems are also in use, and producers call a toll-free number to obtain informa-

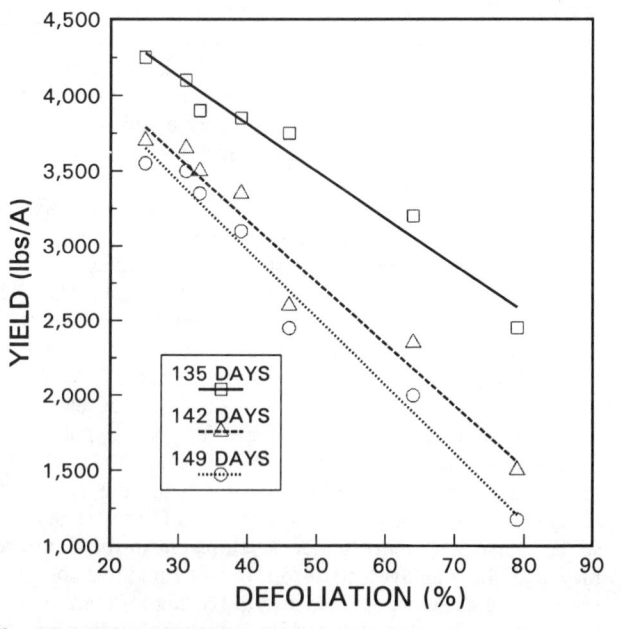

Fig. 10.4. Effect of digging date on the relationship between leaf spot-induced defoliation and pod yield.

tion. Electronic mail systems, in which advisory information is sent from a central mainframe computer directly to the grower and/or county agent, are also being utilized. Local weather (leaf wetness, relative humidity, and temperature) can vary widely over short distances (because of localized showers and irrigation practices). Therefore, some producers may choose to buy their own on-site personal computers to collect and analyze the weather on a field-by-field basis. Since leaf spot diseases are the major foliar disease problem on peanut in the United States, most of the current modeling efforts have been directed toward managing these diseases at a lower cost and with less use of fungicides.

Partial resistance is also well suited for incorporation into an integrated system in which a weather-based spray advisory is used. This integration is accomplished by modifying the rules somewhat to take advantage of the resistance. For example, Southern Runner has been effectively used with a system such as the one described in Box 10.3 by changing the number of days of protection input to the model from 10 to 14.

Even when some type of disease forecasting system is used to determine when fungicide protection is needed, it is still a good idea to scout all fields. Scouting will provide visual backup for the spray advisory system and early warnings of other disease and insect problems.

A comprehensive plan for leaf spot disease management is given in Box 10.4. This plan represents a summary of all the previously mentioned aspects of leaf spot management.

Box 10.3

Typical Spray Advisory Use for Leaf Spot Management on Peanut

One type of advisory system uses an in-field weather station connected to a computer programmed with a leaf spot forecasting model. Ambient temperature, relative humidity, and leaf wetness are recorded hourly each day. When the system is initialized, the crop emergence date is input by using the computer keyboard. The system begins recording weather data from that time, and daily or hourly summaries can be printed on a portable printer (located with the computer in a weatherproof housing) or downloaded to a portable computer in the field or to a desktop computer at the office via cellular phone. The system will accumulate disease index units on the basis of whether a sufficient number of hours of leaf wetness have occurred during times of favorable temperatures. When enough index units have accumulated, a spray advisory is given.

The unit should be checked daily, and the user will see these typical messages:

FORECAST
JULY 25, 1994 3:00 PM
    Spray needed soon
or
    Spray advised
or
    No spray advised

If a spray is advised, the user will be given the post-infection time in hours. A spray should be applied within 72 hours of a "spray advised" message.

Weather data given as part of the daily output include the time at which an ongoing wetness period began, the duration of the wetness period, the average rainfall, and the average temperature.

Other information includes the number of disease index units accumulated during a given wetness period, the cumulative index units, and the number of seasonal index units.

When a fungicide application is made, the date is keyed into the model along with a set number of days of protection (e.g., 10 days). The system will wait until the days of protection have reached zero before beginning to accumulate disease index units again. While the field is under fungicide protection, the system does not need to be checked daily but should be checked every 2–3 days just to be sure everything is operating properly.

Box 10.4

Ten-Point Management Plan for Leaf Spot Diseases

1. Apply a recommended fungicide on a 10- to 14-day calendar schedule or according to a fungicide spray advisory system supported by local research and extension workers. An early leaf spot advisory may or may not adequately control late leaf spot and vice versa.
2. Weather data used to issue spray advisories should be obtained from the weather station located nearest the field to be sprayed.
3. *Option A:* If a spray advisory is used, apply a recommended fungicide within 3 days of the advisory to spray but not within 10 days of the last fungicide application.
   Option B: If a spray advisory is used, apply a recommended fungicide before wet weather if an advisory to spray is expected as a result of that weather.
4. Check spray pressure, nozzles, and output before each application to obtain proper coverage at the recommended fungicide dosage.
5. Scout fields regularly for foliar diseases and revert to a 10- to 14-day calendar spray schedule if any area of the field shows infection levels above 20% (that is, if more than one of five leaflets have a leaf spot). If web blotch or pepper spot becomes a concern, revert to a 10- to 14-day calendar schedule. If peanut rust pustules are found anywhere in the field, spray immediately with a recommended fungicide and continue to spray every 10–14 days.
6. Use recommended leaf spot-resistant cultivars if available and acceptable for markets in the growing area.
7. Utilize rotations that include a minimum of two seasons out of peanut to reduce the amount of inoculum available for initiation of foliar disease epidemics.
8. Reduce the amount of crop refuse by deep plowing or burning.
9. Eliminate volunteer peanut plants.
10. Dig peanuts before defoliation (leaf shedding) exceeds 35% to reduce loss of high-quality pods.

Peanut Rust

If peanut rust, caused by the fungus *Puccinia arachidis* Speg. (*Peridipes arachidis*, proposed), occurs early enough in the season, it can be a destructive disease in the southeastern and southwestern growing areas of the United States. Pod losses in Africa have been reported to be as high as 40–70% from rust and leaf spot, alone or in combination. It is thought that losses from rust may increase faster than those caused by severe leaf spot disease epidemics. Rust may cause pods to mature 2–3 weeks early, seed size may be smaller, and pod detachment at digging may be increased. The oil content of kernels may also be decreased by rust.

This disease is easily diagnosed by the appearance of small pustules (uredinia) on the undersides of leaflets (Plate 78). These pustules are slightly raised areas about the size of a pencil point. Pustules rupture to expose masses of reddish orange urediniospores. Small, yellow flecks are visible about 3 days before the pustules appear. The time from infection with rust spores to the appearance of pustules is thought to be about 10–15 days. These spores are the only source of inoculum for rust infection in peanut, and rust epidemics are made up of several uredinial infection cycles.

Rust appears in fields in distinct focal points of a few to many plants. The disease may then spread rapidly if conditions are favorable. Rust-infected plants are first characterized by the orange pustules, and then leaves die rapidly as the disease progresses. Dead leaves will be rusty brown and later turn black, but unlike leaves affected by leaf spot diseases, they tend to stay on the plant (Plate 79). Free water on the leaves is necessary for spore germination, and infection may occur in 16–24 hours. Localized spore dispersal is favored by light rainfall, and urediniospores can be aerially distributed by wind. Spread of peanut rust spores has been related to daily patterns of variation in relative humidity and wind speed.

The teleomorphic stage, a chestnut brown or cinnamon brown spore that can overseason, does not occur in the United States, and it is thought that *P. arachidis* urediniospores do not survive from season to season. Therefore, it is assumed that peanut rust inoculum is introduced annually from other peanut-producing countries. Tropical storms may play a role in introducing rust inoculum into peanut-growing areas of the United States. Rust pustules were observed on peanut in Texas as early as the first week of July in 1971. Rust reached very high levels that year in south Texas and was widely distributed in the southeastern growing regions. In south Texas, the growing season is longer and peanut may be planted from early March to late July. The peanut crop planted later could be exposed to significant numbers of rust urediniospores if the disease appears early in any given year. Some degree of rust disease management may be achieved by locating later plantings away from earlier ones. In the Georgia-Florida-Alabama area, peanut is planted from late March to May and harvested from late August to November. Rust generally does not occur in that region before August or September, so most of the peanut crop is not exposed to the disease for very long. Peanut rust has not been a problem in the Virginia-Carolina growing area except during years in which hurricanes carried inoculum into the region from the Caribbean area.

The fungicide most frequently used in management of leaf spot diseases, chlorothalonil, has good efficacy against peanut rust, but applications may be necessary at 7- to 10-day intervals if rust is present in the field. The combination of mancozeb and sulfur is also effective against peanut rust. Some sterol-inhibiting fungicides, such as tebuconazole and cyproconazole, have efficacy against peanut rust. Propiconazole is not effective against peanut rust. In fact, use of propiconazole alone late in the season has actually been known to worsen rust epidemics in research plots compared with other fungicides or untreated control plots.

Since the initial rust inoculum is aerially distributed from outside the peanut-production regions of the United States, management practices such as crop rotation and deep plowing are of little value. The most effective tool to combat this disease in areas of the world where it is an endemic problem is crop resistance. Little is known of the rust resistance of peanut cultivars grown in the United States, since many of them have not had any prolonged exposure to peanut rust. The most widely grown cultivar, Florunner, is very susceptible. Two years of observations of the cultivar Southern Runner indicate that it has at least a moderate level of rate-reducing resistance in the field. A management plan for peanut rust is given in Box 10.5.

Web Blotch

Web blotch caused significant yield losses in Oklahoma and Texas during the early 1970s, and outbreaks have been reported in the Virginia-Carolina area. When this disease was first observed in the United States, the predominant cultivars in the Southwest were the highly susceptible Spanish types. Today, these no longer predominate, and resistant runner-types are grown, diminishing the importance of this disease.

Web blotch is caused by *Phoma arachidicola* Marasas, G. D. Pauer, & Boerema. The time from infection to symptom appearance may be as short as 1 week. Symptoms usually observed are irregular brown blotches on the leaflets. However, the type of symptom observed may depend on weather conditions, and a webbed or netlike, tan to bronze pattern may occur on leaflets when temperatures are above the optimum (68°F [20°C]) for disease development. Severe defoliation can occur, particularly if leaves are also infected with late leaf spot. Web blotch apparently requires a longer period of leaf wetness and optimum temperature for infection than early leaf spot. Although web blotch is not currently a serious problem in the United States, it must be considered a potential disease problem. Many new cultivars are becoming available, and little is known of their resistance or susceptibility to this disease. Fungicides used for leaf spot disease management, such as chlorothalonil and mancozeb plus benomyl, have efficacy against the web blotch fungus.

Box 10.5

Peanut Rust Management

- A peanut crop planted late should be located away from one planted earlier.
- Foliar fungicides applied late in the season (from August until harvest) should have good efficacy against peanut rust.
- Use rust-resistant cultivars if available.
- When disease forecasts are used to schedule leaf spot sprays, scout carefully for rust from July until harvest. If any rust foci are noted, begin a 7- to 10-day spray schedule with a fungicide effective against the disease.
- Overhead irrigation should be minimized in fields where rust has been observed.

Other Foliar Fungal Diseases

Other foliar fungal diseases, including Phomopsis leaf spot, Alternaria leaf spot, Leptosphaerulina leaf scorch, and pepper spot, occur on peanut in the United States but have as yet had little effect on production. The causal fungi of these diseases have been periodically isolated from peanut but generally cause only minor cosmetic damage. Under certain conditions, one or more of these diseases may occur and cause serious damage in a peanut field. A severe epidemic was noted by one of the authors in northern Florida in test plots of late leaf spot-resistant breeding lines over 2 years. This occurred when peanut was planted late (late May) and cool wet weather occurred in early September. The leaf spot diseases mentioned above were noted, and either one or a combination of them caused severe defoliation in plots that were not treated with foliar fungicides. Where foliar fungicides were applied for leaf spot diseases, these other disease symptoms were almost nonexistent.

Pepper spot and leaf scorch are considered to be different symptoms caused by the fungus *Leptosphaerulina crassiasca* (Sechet) C. R. Jackson & D. K. Bell. One or both of these symptoms occur annually on peanut plants in all the growing areas. Neither form of the disease is of economic significance, and specific control measures are generally not required.

Potential for Biological Control of Peanut Foliar Diseases

Biological control of late leaf spot disease has been attempted with *Dicyma pulvinata* (Berk. & M. A. Curtis) Arx, a fungus that parasitizes *C. personatum*. Another fungus, *Verticillium lecanii* (A. Zimmerm.) Viégas, has also been reported to parasitize the late leaf spot pathogen and the peanut rust fungus. At least six other fungi have been reported to parasitize the rust fungus. Although these biological control agents occur naturally and have potential for use against foliar diseases of peanut, none of them has yet proved effective as a management tool in the field. At present, biological organisms for use in foliar disease management on peanut are not available and cannot be considered an option for peanut producers. It is hoped that one or more biological control options will become available for use against peanut foliar diseases at some future date. Until then, growers must depend on the integration of cultural practices, the available resistance, and the use of fungicides to manage these diseases.

Selected References

Porter, D. M., Smith, D. H., and Rodríguez-Kábana, R. 1982. Peanut plant diseases. Pages 326-410 in: Peanut Science and Technology. H. E. Pattee and C. T. Young, eds. American Peanut Research and Education Society, Yoakum, TX.

Porter, D. M., Smith, D. H., and Rodríguez-Kábana, R., eds. 1984. Foliar diseases. Pages 5-15 in: Compendium of Peanut Diseases. American Phytopathological Society, St. Paul, MN.

Shokes, F. M., and Smith, D. H. 1990. Integrated systems for management of peanut diseases. Pages 229-238 in: Handbook of Pest Management for Agriculture, vol. 3. D. Pimental, ed. CRC Press, Boca Raton, FL.

Smith, D. H., and Littrell, R. H. 1980. Management of peanut foliar diseases with fungicides. Plant Dis. 64:356-361.

H. A. Melouk
U.S. Department of Agriculture
Agricultural Research Service
Stillwater, Oklahoma

P. A. Backman
Department of Plant Pathology
Auburn University
Auburn, Alabama

CHAPTER ELEVEN

Management of Soilborne Fungal Pathogens

Shortly after planting, a peanut grower may encounter problems, such as reduced germination and poor stand establishment, related to soil environment. Several soil-inhabiting pathogens cause diseases that adversely affect peanut health and productivity throughout the growing areas of the United States. Some of these diseases, such as pod rot complex, crown rot, southern blight, and root knot, occur in all of the peanut-producing areas of the United States. Others are problems only in certain production areas. Blackhull disease, for example, is limited to New Mexico, Oklahoma, and Texas, and Sclerotinia blight is limited to Oklahoma, North Carolina, Texas, and Virginia.

Most of the pathogens that cause these diseases have broad host ranges, including crops and weeds, and are able to live on plant debris and produce durable survival structures that reduce the efficacy of disease-management practices such as crop rotation. Compounding these problems is the shortage of effective chemicals available for use by growers. In this chapter, we discuss the most common and important soilborne disease problems affecting the peanut plant throughout its life, including seed and seedling diseases; Aspergillus crown rot; Rhizoctonia limb, pod, and root rot; stem rot (southern blight); Cylindrocladium black rot; Sclerotinia blight; Verticillium wilt; Pythium pod rot; blackhull; and other minor diseases.

Seed and Seedling Diseases

Causal Organisms

Germinating seed and seedlings of peanut are prone to attack by several pathogenic fungi, including *Rhizopus, Penicillium, Aspergillus, Fusarium, Rhizoctonia,* and *Pythium* species. All these fungi can cause seed and seedling rots. Several are carried on the seed. Problems associated with seed and seedling diseases can be avoided by planting good quality seed with high rates of germination (Chapter 3). Peanut seed and seedlings are highly susceptible to disease during the germination and stand-establishment processes. Several seed protectants are available commercially that will reduce the occurrence of seed and seedling rots.

Typically, even high-quality seed, if untreated with fungicides, provide stands of only 50%. If soils are heavy, cold, or poorly rotated, this same high-quality seed provide even lower rates of emergence. Seedborne fungi appear largely responsible for this problem. The microflora of the seed are often reflective of the soils from which they were harvested. In dry years, seed are heavily colonized with *Aspergillus* spp., while in wet years, the *Rhizopus* fungi are dominant. After planting, the seed and seedlings are attacked by soilborne fungi, particularly *Rhizoctonia, Pythium,* and *Fusarium* species.

> **Typically, even high-quality seed, if untreated with fungicides, provide stands of only 50%.**

Because peanut seed originate in the soil, the stresses to which they are subjected before planting are more severe than those of other crops. The process of drying the harvested pod in the sun for 3 days or longer can result in a loose seed coat (testa) if drying occurs very rapidly or the seed is overdried. Usually, seed are dried with supplemental heat after harvest. If the drying temperature is too high, seed quality is again reduced. Seed are often shelled months before planting, yet fungicides are usually applied only just before the seed are sold to growers to assure that all treated seed will be sold for seed use. Research has indicated that if seed are treated immediately after shelling, seed emergence can be raised 5–10%, probably because wounds that occur during the shelling process are immediately treated with the fungicide and are thus not available for colonization by seed pathogens.

A standard practice in the United States is to scan shelled peanut seed with an electric eye and remove any discolored kernels. The same benefit can be gained by removing discolored kernels by hand as they pass over a conveyor belt. These practices can greatly improve final seed germination rates. Peanut seed harvested from drought-stressed plants usually have reduced vigor. This is reflected in slower rates of emergence and more damage from seedborne and soilborne fungi.

Germination testing and certification provide guarantees to farmers that seed meet minimum standards for germination and cultivar purity and provide assurances that they are not contaminated with noxious pests. In order to meet these germination standards, peanut seed treated with fungicides should typically germinate at least to the 80% level in trials conducted in germination chambers. If this minimum cannot be met, a different seed lot should be selected. Seed lots from different environments often are contaminated with different seed pathogens, so fungicides and fungicide mixtures may perform differently on different seed lots. Therefore, available fungicides or mixtures should be evaluated separately to determine the best product for each seed lot.

Management

Peanut seed lots generally require treatment with fungicides to assure an adequate plant stand in the field. In some countries where insect problems are severe, these fungicides may have to be mixed with insecticides or an insecticide may be applied in the seed furrow. Fungicides are typically applied as dusts. If they are applied in aqueous suspensions, the seed coat is loosened, resulting in untreated, exposed cotyledonary tissue that is usually invaded by pathogens before emergence. Oil-based fungicide formulations have recently been developed that do not loosen the testa and have the additional advantage of not producing dusts during the treatment and planting processes that could expose workers to environmental hazards.

Germination tests of seed lots conducted in growth chambers are good predictors of potential field performance, unless the field has a severe problem with soilborne diseases. Seed treatments, while providing excellent control of seedborne pathogens, provide only limited control of the soilborne fungi that attack seed and seedlings. If these are known to be a problem, in-furrow applications of granular or spray fungicides should be used. Only a low percentage of fields in the United States are treated with in-furrow fungicides.

In years with well-distributed rainfall, the pathogens most often found on the harvested seed are *Rhizopus* spp., most commonly *R. stolonifer* (Ehrenb.:Fr.) Vuill. and *R. arrhizus* A. Fischer. Control is achieved with DCNA (2,6-dichloro-4-nitroaniline), which is mixed with captan, thiram, or maneb to expand the spectrum of control, although these three fungicides afford relatively good control against *Rhizopus* spp. In dry years, *Aspergillus* spp., especially *A. flavus* and *A. niger*, dominate in the soil and on the seed. Captan, maneb, and thiram combinations usually are quite effective in controlling this group of fungi as well as *Penicillium* and *Fusarium* species when they are present. Carboxin and/or PCNB (pentachloronitrobenzene) can be added to the combinations to provide some protection against soilborne *Rhizoctonia*. Seed protectants are usually applied at 4 ounces per 100 pounds (220 grams per 100 kilograms) of seed.

Cultural measures can reduce the impact of seed and seedling diseases on stand establishment. Peanuts should be planted at 4 inches (10 centimeters) only if soil temperatures exceed 65°F (18°C). If soils are cool and wet, planting depths can be reduced to as little as 1–2 inches (3–5 centimeters). If conditions are hot and dry, planting depths can be as much as 5.5 inches (15 centimeters) to assure seed placement in moist soil. The more quickly the seedling emerges, the greater its probability of escaping disease and the less damage will be exacted by seed and seedling pathogens. Rotations can also reduce the impact of soilborne fungi and should be used particularly if stand establishment has been a problem and quality seed have been used.

Aspergillus Crown Rot

Aspergillus crown rot is caused by *Aspergillus niger* Tiegh. The disease can be found on peanut in all major growing areas of the world, although its appearance is generally sporadic and unpredictable.

Symptoms

Peanut seedlings and young plants are very susceptible to *A. niger* and suffer sudden loss of leaf turgidity and wilting. Crown tissues of infected plants become swollen or enlarged, and the tissues become brittle and corky. Decay and shredding of the taproot generally follows death of the affected plant. Infected plant parts are profusely covered with black masses of mycelia and conidia (spores), which resemble black powder (Plate 80). On Spanish peanut, a major symptom on an infected older plant is the dead central stem, which is often broken at the soil line.

Disease Initiation, Development, and Spread

A. niger is ubiquitous in soils throughout the world. This fungus colonizes organic matter, where it produces abundant, large, black, conidial heads that reach 0.02 inch (0.5 mm) in diameter. The fungus, along with *A. flavus* Link:Fr., dominates drought-affected soils (Chapter 13), and disease is more severe during dry periods. The fungus attacks the germinating seed in the cotyledonary and hypocotyl regions shortly after germination (Plate 81). Disease progresses rapidly, and most young plants die within 30–40 days after infection. The population levels of the pathogen found in very wet soils are much lower than those found in dry soils. The fungus often is prevalent in soils low in organic matter and is also more prevalent in soils continually cropped to peanuts than in soil of rotated fields.

Disease Losses

All commonly grown peanut cultivars are susceptible to the crown rot fungus. Cultivars with a bunching growth habit usually are less susceptible than runner types. *A. niger* is seedborne, and the infestation levels in seed lots often exceed 90%. Such infected seed produce a high percentage of infected plants. Stand losses caused by the crown rot fungus are variable and difficult to assess. In highly infested fields, they can reach 50%, but they usually vary from a trace to 1%.

Management

Losses caused by crown rot can not be eliminated from peanut fields. However, they can be minimized by 1) planting seeds with high germination and vigor, 2) planting seeds treated with chemical protectants, 3) using timely irrigation to avoid extreme drying of soils, and 4) avoiding injury or damage to seedlings from pesticides.

Rhizoctonia Limb, Pod, and Root Rot

The primary pathogen responsible for the *Rhizoctonia* complex of diseases in peanut is *Rhizoctonia solani* Kühn. This fungus can cause disease on all parts of the peanut plant. Significant damage has been reported from Australia, Egypt, India, Africa, the Caribbean, and the United States. Damage to peanut pods and pegs has long been recognized, but recently research has indicated significant losses resulting from damage to taproots in the form of dry, sunken cankers and to peanut branches in the form of girdling cankers and leaf rot.

R. solani is found wherever peanut is grown and can be easily isolated from either soil or infected plant parts. The fungus produces numerous dark brown to black survival structures (sclerotia) in or on the infected plant.

Symptoms

Infection by *R. solani* may cause seedling damping-off, or it may cause dry, sunken cankers on the hypocotyl (the portion of the plant between the taproot and the aboveground parts) and taproot. Later in the growing season, vines may become infected when the canopy is well established and there is a warm, moist subcanopy environment. This usually occurs when plants are in the early reproductive stages, and the first symptoms develop on limbs that are in contact with the soil or that have been damaged by equipment. As the disease progresses, the cankers expand both laterally and longitudinally until the vine is girdled (Plates 82 and 83). At the same time, particularly in irrigated fields or in peanuts grown under conditions of frequent rainfall, the fungus causes a blight of leaves in the lower canopy (Plate 84). These dark brown leaves are matted together with mycelium of *R. solani* and are diagnostic for the disease. Limbs in advanced stages of disease development will appear tattered. Most of the cortical tissue is gone, and only the vascular bundles remain. Sclerotia are easily seen in the remaining tissues. The pegs are frequently destroyed, causing a loss of the pods in the ground.

Pod rot by *R. solani* occurs after direct penetration of the pod by the fungus. When first attacked, the pod displays brown to dark brown cankers on its surface. As the disease progresses, the developing seed is colonized, killed, and then digested to the point that there are frequently dry mummies remaining within the blackened shell. Lesser degrees of damage result in discolored kernels that reduce the quality and price of the harvested product.

Damage from *Rhizoctonia* may be confused with damage caused by other soilborne pathogens such as *Sclerotium rolfsii*, *Sclerotinia minor*, and *Pythium* spp., which are discussed in later sections of this chapter. Close examination by the observer for signs of the fungus is helpful in sorting out the various disease problems. For example, pod rot caused by *R. solani* results in dry rot of kernels, while *Pythium* spp. produce a wet rot.

Disease Initiation, Development, and Spread

R. solani can survive in field soils for extended periods of time. This prolonged survival is the result of its wide host range and ability to use crop and weed debris as a food base. Sclerotia, produced in infected host tissues, serve as the overwintering and initial inoculum. Sclerotia germinate when stimulated by secretions from a susceptible host or by the presence of organic matter that can initiate the infection process.

R. solani can be spread by farm workers, machinery, water, and windblown plant debris and soil. Box 11.1 summarizes factors that favor development of *Rhizoctonia* diseases.

Disease Losses

Rhizoctonia-induced diseases of peanut are usually found as a complex of soilborne diseases, and the severity of each and the individual contributions to disease-induced losses are difficult to determine. Typical evaluation systems for *Rhizoctonia* damage include the percentage of limbs (vines) damaged, the percentage of pods discolored (dry rot), and/or the percentage of root surface area cankered. A recent study indicated that the losses induced by the root rot phase of the complex, in which 25% of root surface area was damaged, caused losses of more than 775 pounds per acre (879 kilograms per hectare).

One of the best estimates of the extent of losses can be made from the evaluation of results from chemical disease-control trials. Typically, where the sterol-inhibiting fungicides tebuconazole or diniconazole were applied, yields improved by 705–1,410 pounds per acre (800–1,600 kilograms per hectare) compared with plots receiving chlorothalonil for control of peanut leaf spot. With diniconazole, yields increased an average of 1,145 pounds per acre (1,300 kilograms per hectare) in a regional experimental use trial. With tebuconazole, yields improved 795 pounds per acre (900 kilograms per hectare) in 29 trials conducted at locations with low to moderate soilborne disease severity and 1,320 pounds per acre (1,500 kilograms per hectare) on 12 sites with moderate to severe soilborne disease severity. In each case, control of southern stem rot contributed to the yield response, but investigators frequently reported control of *Rhizoctonia* to be nearly equal in importance.

Management

Disease management should be based on reducing the population of the pathogen through crop rotation and modifying cultural practices. Rotations with grass crops, such as small grains, corn, sorghum, and forage grasses, are most effective in delaying the buildup of *Rhizoctonia* populations. However, rotations to reduce the severity of an already serious problem may take 3–5 years to decrease pathogen populations below damaging levels. In the United States, the *Rhizoctonia* complex has been increasing in severity during the late 1980s

Box 11.1

Factors that Favor Development of the *Rhizoctonia* Complex

- Short or nonexistent rotation with peanuts or other legumes in the previous 2 years
- Early planting in cool soils, which slows peanut emergence
- Dense vine growth brought on by irrigation or frequent rainfall
- Damage to vines resulting from the use of equipment in the peanut field

and early 1990s. One reason is the use of shorter rotations with only 1 year between peanut crops for the large majority of farmers. A second is the extensive use of irrigation, which can cause plant growth to be excessive and thus produce a subcanopy environment that is conducive to disease development. In all fields, applications of water should be made only when necessary to replenish soil moisture.

Some state recommendations include the application of gypsum to fields with histories of peg and pod rot problems to correct calcium deficiencies thought to predispose peanuts to this phase of the disease. Adequate calcium levels within the pegging zone are important for healthy pod development.

> **Rotations with grass crops, such as small grains, corn, sorghum, and forage grasses, are most effective in delaying the buildup of *Rhizoctonia* populations.**

Applications of growth hormones have been used in the past to reduce vine development, which contributes to better aeration and lowers the buildup of humidity within the plant canopy, thus reducing disease severity. Practices that reduce vine damage from the passage of equipment through the fields are also reported to result in lower levels of limb and pod rot. Low levels of resistance have been reported in some cultivars, but disease still develops to levels requiring management. Sterol-inhibiting fungicides such as tebuconazole can provide up to 50% reduction in the number of lesions on the lateral branches and reduction in the expansion of existing lesions.

Stem Rot

One of the most widely recognized diseases of peanut is stem rot, caused by the soilborne fungus *Sclerotium rolfsii* Sacc. Other names for this disease include white mold, southern blight, and Sclerotium rot. This fungus attacks and kills the crown of the plant, removing whole plant units and the pods they are producing. The disease is found in virtually all peanut-growing regions of the world. In the southeastern peanut-producing region of the United States, it is often considered to be responsible for the greatest losses of those caused by soilborne pathogens.

S. rolfsii can generally be classified as a necrotrophic fungus; that is, it kills the tissue in advance of the growing mycelium, which then digests the plant material. The fungus produces large amounts of oxalic acid, a phytotoxin, which kills the tissue in advance of the mycelium and in the early stages of disease development, causes necrosis (dead tissue) and chlorosis (yellowing) in foliage. Purple seed stain occurs as a result of oxalic acid damage from fungal growth in and around the developing pod.

Symptoms

The first symptoms of stem rot are usually yellowing and wilting of a branch or, if the main stem becomes infected, of the whole plant. Leaves turn dark brown and are sometimes shed prematurely. Sheaths of white mycelium can be seen at or near the soil line around affected plant parts (Plate 85). Mycelial growth is rapid under warm, moist conditions and

quickly spreads to other branches and plants. Sclerotia, 0.2–0.8 inches (0.5–2.0 mm) in diameter, are produced abundantly, both on plant parts and on the soil surface (Plate 86). They are initially white but later turn dark brown. Lesions produced on branches and pegs are initially light brown and become dark brown as the disease progresses. Infected pods are usually rotted and may occur on plants without visible symptoms on the aboveground parts. Oxalic acid produced by the fungus may cause a purple discoloration of seeds in mildly affected pods.

Disease Initiation, Development, and Spread

S. rolfsii has a broad host range of more than 200 plant species. Sclerotia are produced on many infected crops and weeds, and even debris from nonhosts may serve as an organic food base for attack of subsequent crops. Disease severity is a function of sclerotia population levels, as has been proved in sugar beets (*Beta vulgaris*); but in peanut, the population levels required to produce severe disease are so low that no effective systems have been developed to accurately and quickly assess them.

This fungus has a high demand for oxygen. Therefore, the overwintering sclerotia activated are usually those that occur in the upper regions of the soil. However, in soils that crack deeply when dry (e.g., vertisols), oxygen can penetrate the soil profile, activating sclerotia well below the surface and thus allowing pod and root rots that would not occur if the same soils were wet. Another factor in sclerotial activation is the midseason loss of leaves that occurs through natural senescence and disease-induced defoliation. This decomposing tissue releases alcohols and other volatiles that stimulate sclerotial germination. Such dead tissue then serves as a food base once the germination process has begun. These organic bridges facilitate plant-to-plant spread as does debris from the previous season's host and nonhost weed and crop plants.

Unfortunately, control of peanut leaf spot can indirectly increase the severity of other diseases. Maintaining a complete canopy allows a moist subcanopy environment that is conducive to the development of stem rot and diseases caused by *Rhizoctonia*. Also, if the leaves are healthy and the canopy intact, less fungicide from leaf spot control activities reaches the soil surface and less nontarget disease control will result. However, the results of not controlling leaf spot diseases would be far worse.

Studies indicate that deeply buried sclerotia survive a year or less, while those near the soil surface can remain viable for 3–4 years. Stem rot develops when sclerotia germinate close to plant structures lying in or on the soil surface. If cultivation for weeds moves soil onto the limbs or crown of the plant, disease is enhanced because more sclerotia occur near susceptible plant parts. Similarly, plants with a more erect growth habit, and thus with less soil contact by plant parts, will suffer less damage.

Disease Losses

Losses from stem rot average 7–10% annually in the southeastern United States. For the southwestern and Atlantic states, losses are usually less than one-half these levels. Disease damage, usually represented by numbers of dead plants, can be easily seen during warm, wet weather.

Since damage from any single infection focus usually kills several plants, rating systems have been developed that count infection loci per length of row. One such system makes the assumption that any single infection locus (site) will not spread beyond 6 inches (15 centimeters) from its source. An area of dead plants of less than 12 inches (30 centimeters) row length is then assumed to originate from a single source. For areas between 12 and 24 inches (30 and 60 centimeters) of row, there are probably two loci. A strong negative correlation between numbers of disease loci per unit row length and yield has been reported. The relationship between disease loci and yield differs from year to year, primarily because damage to pegs and pods is affected by the environment. Losses of 9–45 pounds per acre (10–50 kilograms per hectare) for each infection locus in a 98-foot (30-meter) length of row for runner peanuts were reported by this method, which illustrates the variability of losses that may occur. Counts of disease loci are usually made in the late pod fill stage (100–120 days after planting). If there are more than three to four loci per 98 feet (30 meters), then control measures are warranted.

During dry periods, *S. rolfsii* causes more peg and pod damage because it is more active in the pod zone. When peg and pod rot damage are particularly evident, some investigators prefer to enumerate numbers of disease loci (defined as damaged plants with signs of the fungus) in the field after plants are dug for harvest. Disease loci are the same size as those defined previously.

Management

Management of stem rot begins with prevention of inoculum buildup. Deep plowing after harvest to bury crop debris aids in decomposition of the organic matter used by the fungus as a saprophytic food base and a bridge between plants. Furthermore, deeply buried sclerotia have much shorter survival times than those near the soil surface. Cultivation for weed control will reduce disease buildup if the farmer is careful to prevent the movement of soil onto the plant crown and lateral stems. Pesticides must be carefully selected to reduce the severity of stem rot. For instance, the leaf spot fungicide benomyl (Chapter 10) will increase the severity of stem rot by reducing levels of the natural biological control fungi *Trichoderma* spp. However, the dinitro herbicides (Chapter 7) and the insecticides ethoprop, fensulfothion, and chlorpyrifos are all reported to suppress the severity of stem rot and have been marketed for this purpose.

Rotations can be effective in preventing development of severe infestations. Particularly effective are 1-year rotations with corn or grain sorghum. Rotations of 2–4 years are required if a severe situation has developed. Fallowing will work only if the land is kept nearly weed free. This is necessary because of the extensive host range of *S. rolfsii*.

Chemical control with PCNB (quintozene) or carboxin is practiced in the United States. Typically these products offer 40–60% control. For PCNB, rates of 4.85–9.7 pounds per acre (5.5–11 kilograms per hectare) of active ingredient (a.i.) are delivered in a 4.0- to 12.0-inch (10- to 30-centimeter) band centered over the row. The preferred formulation is a 10% granule to facilitate canopy penetration, although some success has been reported with directed sprays. Carboxin is usually applied as a granule at 1 pound a.i. per acre (1.1 kilogram a.i. per hectare). Liquid sprays of carboxin, although more effective in control of the disease, cause chlorosis of the foliage and are not recommended. Both products can be delivered through irrigation water. Also, sterol-inhibiting fungicides such as tebuconazole provide good control for both southern blight and leaf spot (Chapter 10).

Combination granules of PCNB plus chlorpyrifos, fensulfothion, or ethoprop have been marketed in the United States. Chlorpyriphos applied as a 15% granule at 2.0 pounds a.i. per acre (2.2 kilograms a.i. per hectare) for control of the lesser cornstalk borer, *Elasmopalpus lignosellus* (Chapter 8), reduces stem rot an average of 40% if applications are made during the pegging period.

Until recently, only low levels of tolerance to stem rot had been reported in peanut cultivars or breeding lines. Generally, cultivars with more erect growth habits had less disease than cultivars with spreading growth habits (and thus more soil contact). However, according to a recent report, Southern Runner was found to have 65% less stem rot than Florunner over the course of a 3-year study. Both cultivars have spreading growth habits, although Southern Runner does not have an apically dominant central stem like Florunner.

Cylindrocladium Black Rot

Cylindrocladium black rot (CBR) was first reported in Georgia in 1965. Subsequently, the disease spread to several other locations in the southeastern United States and the Virginia-Carolina peanut-growing region, where it is considered economically damaging. The disease has also been reported in Japan, Australia, and India.

CBR is caused by the fungus *Cylindrocladium crotalariae* (C. A. Loos) D. K. Bell & Sobers, which produces microsclerotia in infected roots and rhizobium nodules. These microsclerotia are structures that resist adverse conditions and serve as a survival units. Microsclerotia vary in size, ranging from 53 to 88 micrometers, and are readily visible in infected tissue without staining.

Symptoms

The first symptoms of CBR are wilting and chlorosis of infected leaves and stems (Plate 87). Usually the central stem is the first to show symptoms, which are easy to see even from a distance. As the disease progresses, the lateral branches also will show symptoms. Occasionally, the whole plant will become chlorotic and stunted. All underground plant parts may develop symptoms of CBR. Hypocotyls, roots, and pods become black and necrotic. When CBR is seen for the first time in a field, the infested area develops in one or more localized spots, which sometimes measure 33–99 feet (10–30 meters) in diameter. A sign of CBR that can be used for diagnosis is the occurrence of small, reddish orange perithecia (sexual structures for reproduction of the fungus) in dense clusters on stems, pegs, and pods (Plate 88). These structures form only on tissues within a few millimeters above or below the soil surface during periods of wetness and high humidity. If these symptoms are lacking, a positive diagnosis of CBR requires a laboratory assay.

Disease Initiation, Development, and Spread

Microsclerotia of *C. crotalariae* are produced extensively in infected roots. As the infected tissue decomposes, these microsclerotia are released into the soil and serve as the primary source of inoculum for disease initiation. Microsclerotia are not uniformly distributed in infested fields, so infected plants tend to occur in clusters. Root exudates may stimulate the microsclerotia to germinate and infect the root tip.

Microsclerotia overwinter in the soil, providing the inoculum for the next season. Winter soil temperatures below 40°F (5°C) for extended periods will reduce viable populations, as will short periods during which the soil freezes. Disease severity is directly related to microsclerotia population levels at planting. Soil temperatures near 75°F (25°C) with soils near field capacity result in optimal disease development. Reports from the Virginia-Carolina region indicate that if June and/or July are wet followed by a dry August and September, then disease severity will be near maximum. The early wet conditions allow infections to occur and root rot to set in, while the later drought conditions increase severity by limiting the number of functional roots. Root-knot and ring nematodes (Chapter 12) have been reported to potentiate disease severity, perhaps by reducing root function, which increases plant stress, and by providing wound infection sites.

The pathogen may spread via infested soil and plant debris attached to tillage and harvest machinery. Wind and water can contribute to long-range dissemination, and farm workers can carry the pathogen from one field to another.

Disease Losses

CBR destroys all underground plant parts and causes major yield losses in infested areas. Diseased plants have a characteristic infrared signature that has allowed disease surveys to be conducted by aerial photography and remote sensing. Depending on the peanut cultivar planted, losses in affected areas of a field may range from less than 1% to more than 50%.

Management

Current recommendations for management of CBR include crop rotations with nonlegumes such as corn, sorghum, or forage grasses, control of nematodes, and the use of resistant cultivars. Rotations should be 3 years or longer, unless low soil temperatures during the winter have greatly reduced viable microsclerotia populations.

Spanish cultivars are most resistant to CBR; Valencia cultivars are least resistant; and Virginia cultivars are intermediate. Resistant peanut cultivars include NC 8C and NC 10C. Resistance in these cultivars can be broken if they are planted in soils with large populations of microsclerotia. Fumigants provide excellent control when applied to soil (at soil temperatures above 60°F [15°C]) in a band before planting. Yields of resistant cultivars are also increased with fumigation.

> **Spanish cultivars are most resistant to CBR; Valencia cultivars are least resistant; and Virginia cultivars are intermediate. Resistant peanut cultivars include NC 8C and NC 10C.**

Tillage practices, such as removal and destruction of peanut hay and postponement of disking soil until spring, may contribute to reduced viability of microsclerotia. Washing field equipment can reduce spread of the pathogen between fields.

Sclerotinia Blight

Sclerotinia blight of peanut, caused by the fungus *Sclerotinia minor* Jagger, has become one of the major limiting factors in peanut production in both the Southwest (Oklahoma and Texas) and the East (Virginia and North Carolina). The disease was first found on peanut in Virginia in 1972 and has since spread to the other peanut-producing areas of the United States.

Symptoms

In fields with a history of Sclerotinia blight, the disease usually appears shortly after row lapping. The first symptoms occur on the foliage at the plant top, which turns light green to yellow and loses turgor (called flagging or wilting). Examination of the lower canopy early in the morning usually reveals the presence of fluffy, cottony mycelia on the main stem or lateral branches and other affected plant parts, including the taproot near the soil line (Plate 89). Within 3–4 days, these mycelia will mat, and black, dark bodies called sclerotia form on the outside and inside of infected tissues (Plate 90). These sclerotia can also be found on or in the pods produced on infected plants (Plate 91). The sclerotia are irregular in shape and measure about 0.04–0.12 inch (1–3 millimeters). Lesions caused by the infection of stems and branches are tan to brown. Severely infected plants turn dark brown to black and die. Shredding of infected stems, branches, and pegs is a characteristic sign of this disease (Plate 92).

Disease Initiation, Development, and Spread

Sclerotia of *S. minor* can persist in field soils for 4–5 years in the absence of peanut. One sclerotium in 3.5 ounces (100 grams) of soil is reported to be sufficient to initiate the disease, which may become severe in peanut when environmental conditions are conducive to its development. Infection with *S. minor* is favored by cool temperatures (64–68°F [18–20°C]), moist soils (resulting from high rainfall or excessive irrigation), and high relative humidity (95–100%) in the lower peanut canopy. Excessive vine growth that provides a dense canopy may result in more rapid disease development.

It has been shown that the sclerotia of *S. minor* remain viable after passage through the digestive tract of a crossbred

Box 11.2

Tips for Managing Sclerotinia Blight

- Avoid planting peanuts in fields with a history of the disease.
- Plant *Sclerotinia*-free seed or seed treated with chemical protectants.
- Plant peanut cultivars with moderate resistance to the disease, such as VA 81B, VA 93B, AD 1, Southwest Runner, and Tamspan 90.
- Avoid or minimize injury of peanut vines by machinery.
- Early planting and low seeding rate can provide some disease suppression.
- Avoid excessive irrigation at any time during the growing season when cool temperatures prevail.
- Rotate peanut with a nonleguminous crop such as corn or sorghum.
- Use EPA-approved fungicides and follow the manufacturers' directions.

heifer; therefore, feeding peanut hay infested with sclerotia of *S. minor* to ruminant animals might enhance the spread of Sclerotinia blight from infested fields to clean fields. Another important mode of potential disease spread is the movement of infested peanut hay to clean fields. Also, peanut seed harvested from *Sclerotinia*-infected plants have the potential to be contaminated or infected with the fungus. Therefore, planting contaminated seed could serve as a vehicle for the spread of the pathogen. The Sclerotinia blight fungus can also be spread by farm workers, machinery, water, and soil blown by the wind.

Disease Losses and Management

Yield losses of 10% are common with this disease (Plate 93). In cases of severe infection and rapid disease development, however, up to 50% of the pods can be lost.

Measures for effective control of this disease are not available. However, there are means to minimize the losses (Box 11.2). Runner- and Virginia-type peanuts are most susceptible and sustain the most damage from Sclerotinia blight. A few resistant peanut cultivars are now available (Chapter 4).

Verticillium Wilt

Verticillium wilt of peanut, caused by the fungus *Verticillium dahliae* Kleb., was first observed in the United States in New Mexico in 1958. Verticillium wilt is found in all areas of peanut production in the United States, but its economic importance is limited to New Mexico and Oklahoma.

Symptoms

The first symptoms of Verticillium wilt appear about the time of flowering, but older plants may also be infected. Early symptoms consist of marginal chlorosis of leaflets, decrease in leaf turgidity, and curling of leaves (Plate 94). As the disease progresses, the whole plant yellows and leaf necrosis and defoliation occur. The final symptoms are stunting, wilting, and dehydration of infected plants. These symptoms intensify under conditions of moisture stress and day temperatures exceeding 78°F (26°C). Infected plants usually mature early. Brown discoloration of the vascular (water-conducting) system in crowns and stems of infected plants usually occurs, and this discoloration also occurs in roots and petioles when wilt symptoms are severe (Plate 95). The wilt fungus produces abundant microsclerotia that resemble coarse, black powder and that can be seen without the aid of a magnifying lens in infected plant parts including stems, roots, pegs, and pods.

Disease Initiation, Development, and Spread

V. dahliae survives in soil by forming black microsclerotia that can resist adverse environmental conditions for several years in the absence of peanuts. These microsclerotia remain dormant until peanut root exudates stimulate them to germinate. The fungus infects the plant through the root system and then develops rapidly and systemically, spreading throughout the vascular system to all plant parts.

The pathogen spreads by movement of farm machinery from infested to noninfested fields. Spread of the pathogen between fields can also occur by wind and the movement of infested, waterborne soil and plant tissue. Another possible mode of spread of the wilt fungus is by infected peanut seed. The pathogen infects a wide range of hosts, including some weeds, and thus is maintained in soil in the absence of the peanut plant.

Disease Losses

In Argentina, a reduction in peanut yield of 60% has been reported from Verticillium wilt. Over the years in the United States, disease losses have been moderate. For example, in 1980, an estimated reduction in peanut yield of 2% occurred. Valencia and Spanish peanuts are more susceptible to Verticillium wilt than Virginia market types.

Management

Verticillium wilt in peanut can be managed in several ways. 1) Irrigate infested fields more frequently to reduce water stress of infected plants and to allow plants to survive and mature until digging. 2) Plant clean, *Verticillium*-free seed. 3) Control weeds. 4) Rotate peanuts with nonsusceptible host crops, such as corn and sorghum, and avoid cotton and crops belonging to the tomato family, such as potato. 5) Remove and burn infected crop residues to reduce the buildup of the pathogen in field soil. 6) Fumigate the soil with approved chemicals, and follow the label directions.

Pythium Pod Rot

Pod rot or pod breakdown of peanut is a common disease problem in all of the peanut-growing areas of the world. In the United States, the disease has been endemic for at least 40 years.

The fungus *Pythium myriotylum* Drechs. has been implicated as the major cause of pod rot of peanut in North Carolina, Virginia, Georgia, and Oklahoma.

Symptoms

In general, no foliar symptoms are associated with mild to moderate pod-rotting activity. However, when severe pod rot is present, the roots of affected plants may exhibit some deterioration that can result in sudden wilt. Taproots and fibrous roots are susceptible to decay by *P. myriotylum*, and root tissues turn dark brown to black when infected. Leaflets of wilted plants become chlorotic (light green). Nodules can also be infected with *P. myriotylum* and ultimately turn brown to black.

Under moist soil conditions, infected pegs and pods appear water soaked. Infected immature and mature pods exhibit degrees of brown discoloration on part or all of the affected fruit (Plate 96). Complete disintegration of the whole pod is also possible. No visible signs of the pathogen are found on plants with rotted pods.

Disease Initiation, Development, and Spread

P. myriotylum can persist in soils as a saprophyte (an organism that uses dead material for food) for long periods of time. Oospores (sexually produced reproductive bodies) are the primary long-term, soilborne, survival structures of the fungus and are capable of directly penetrating peanut pods and causing infection. Infection by *P. myriotylum* and the development of pod rot occur more rapidly under continuous wet conditions and soil temperatures of 86–94°F (30–34°C).

Soilborne mites and root-knot nematodes (Chapter 12) have been frequently shown to increase the incidence and severity of pod rot caused by *P. myriotylum*. Also, injury to pods from the feeding activities of soil-inhabiting insects and small animals may provide entry points for infection by pod-rotting microorganisms.

Disease Losses

Losses caused by Pythium pod rot range from a trace to more than 50%. It is difficult to assess the damage resulting from infection with *P. myriotylum* alone because other soil microorganisms, such as *Rhizoctonia solani* and *Fusarium solani*, are involved in the pod rot complex.

Management

Because of the complex nature of peanut pod rot, it is a difficult problem to manage successfully. In general, Spanish peanut cultivars are more resistant to pod rot than Virginia or runner market types.

Some soil fumigants have been successful in reducing losses caused by pod rot. Significant suppression of peanut pod rot has been obtained in the southeastern United States and Texas with the application of high rates of calcium in the form of gypsum (calcium sulfate). However, the addition of gypsum did not reduce pod rot severity in Oklahoma or Israel.

Crop rotation has not been effective in reducing peanut pod rot incidence and severity. However, fields with a history of pod rot that are fallow for 2 years have significantly less pod rot than fields cropped to peanut continuously for several years.

Blackhull

Blackhull is caused by the fungus *Thielaviopsis basicola* (Berk. & Broome) Ferraris. The occurrence of the disease in the United States is limited to the peanut-growing areas of New Mexico, where it is severe on the Valencia and Spanish types. The disease is rare on runner-type peanuts.

Symptoms

As the name implies, blackhull is characterized by the appearance of small, scattered, blackened patches on the surface of the pod, which can coalesce to form larger patches (Plate 97). This discoloration is caused by the large masses of black chlamydospores (asexual survival spores) formed within the hull tissues. Black discoloration (black lesions) can also occur on severely infected pegs, causing the pegs to rot and resulting in pod loss in severely infected fields. No aboveground symptoms are associated with this disease.

Disease Initiation, Development, and Spread

Chlamydospores of *T. basicola* (Plate 98) persist in field soils for long periods. The fungus grows well on plant debris in soils, which contributes to its long survival. Chlamydospores germinate, forming infection hyphae that attack pegs and pods. The mycelium of *T. basicola* grows rapidly on infected tissues, and abundant chlamydospores are produced from this mycelium. The disease develops best under conditions of high soil pH (above 7.0), excessive soil moisture, vigorous vine growth, and cool soil during pod development and maturation.

Infected pods and other plant parts left in the field after harvest contribute to survival of the pathogen into the next growing season and to the spread of the disease by wind. Crop sequences of peanut following cotton, peanut, or sweet potato have been found to increase blackhull severity in New Mexico.

Disease Losses

In heavily infected peanut fields, up to 80% of the pods can be lost at harvest. Also, any amount of black discoloration of the peanut pod can reduce the crop's value if it is sold raw or roasted in the shell.

Management

No resistance to *T. basicola* is available in the peanut cultivars planted in the United States or other countries. Rotation of peanut with nonleguminous crops such as small grains, corn, and sorghum may alleviate peanut losses caused by blackhull disease. Also, application of benomyl in the planting furrow may decrease infections by *T. basicola*.

Other Diseases

Soilborne fungal diseases of minor importance that can cause sporadic problems in peanut production in the United States include charcoal rot, caused by *Macrophomina phaseolina* (Tassi) Goidanich; Diplodia collar rot, caused by *Diplodia gossypina* Cooke; Botrytis blight, caused by *Botrytis cinerea* Pers.:Fr.; Phymatotrichum root rot, caused by *Phymatotrichum omnivorum* Duggar; and Olpidium root discoloration, caused by *Olpidium brassicae* (Woronin) P. A. Dang. A bacterium, *Pseudomonas solanacearum*, causes bacterial wilt.

Selected References

Aycock, R. 1966. Stem rot and other diseases caused by *Sclerotium rolfsii*. N.C. Agric. Exp. Stn. Tech. Bull. 174.

Backman, P. A., Rodríguez-Kábana, R., and Williams, J. C. 1975. The effect of peanut leafspot fungicides on the nontarget pathogen, *Sclerotium rolfsii*. Phytopathology 65:773-776.

Bailey, J. E., and Matyac, C. A. 1989. A decision model for use of fumigation and resistance to control Cylindrocladium black rot of peanuts. Plant Dis. 73:323-326.

Bell, D. K., and Sumner, D. R. 1984. *Rhizoctonia* diseases. Pages 23-25 in: Compendium of Peanut Diseases. D. M. Porter, D. H. Smith, and R. Rodríguez-Kábana, eds. American Phytopathological Society, St. Paul, MN.

Beute, M. K., and Rodríguez-Kábana, R. 1979. Effect of wetting and the presence of peanut tissues on germination of sclerotia of *Sclerotium rolfsii* produced in soil. Phytopathology 69:869-872.

Black, M. C., and Beute, M. K. 1984. Relationships among inoculum density, microsclerotium size, and inoculum efficiency of *Cylindrocladium crotalariae* causing root rot on peanuts. Phytopathology 74:1128-1132.

Brenneman, T. B., and Sumner, D. R. 1989. Effects of chemigated and conventionally sprayed tebuconazole and tractor traffic on peanut diseases and pod yields. Plant Dis. 73:843-846.

Diomonde, M., and Beute, M. K. 1981. Effects of *Meloidogyne hapla* and *Macroposthonia ornata* on Cylindrocladium black rot of peanut. Phytopathology 71:491-496.

Garren, K. H. 1959. The stem rot of peanuts and its control. Va. Agric. Exp. Stn. Bull. 144.

Hau, F. C., Campbell, C. L., and Beute, M. K. 1982. Inoculum distribution and sampling methods for *Cylindrocladium crotalariae* in a peanut field. Plant Dis. 66:568-571.

Parmeter, J. R., Jr., Sherwood, R. T., and Platt, W. D. 1969. Anastomosis grouping among isolates of *Thanatephorus cucumeris*. Phytopathology 59:1270-1278.

Phipps, P. M. 1990. Control of Cylindrocladium black rot of peanut with soil fumigants having methyl isothiocyanate as the active ingredient. Plant Dis. 74:438-441.

Phipps, P. M., and Beute, M. K. 1979. Population dynamics of *Cylindrocladium crotalariae* microsclerotia in naturally infested soil. Phytopathology 69:240-243.

Rodríguez-Kábana, R., Backman, P. A., and Williams, J. C. 1975. Determination of yield losses to *Sclerotium rolfsii* in peanut fields. Plant Dis. Rep. 59:855-858.

Turner, J. T., and Backman, P. A. 1988. Severity, distribution, and losses from taproot cankers caused by *Rhizoctonia solani* in peanuts. Peanut Sci. 15:73-75.

CHAPTER TWELVE

D. W. Dickson
Department of Entomology and Nematology
University of Florida, Gainesville

H. A. Melouk
U.S. Department of Agriculture
Agricultural Research Service
Stillwater, Oklahoma

Management of Nematode Pests

Nematodes are tiny worms with long, cylindrical, unsegmented bodies that attack both plant and animal species. When nematodes attack plants, they cause damage mostly to roots and underground plant parts. The general life cycle of a nematode includes egg, juvenile, and adult stages.

In Chapter 11, the interaction between some nematodes and soilborne pathogens was briefly discussed. This chapter presents an expanded discussion of the nematodes that affect peanut health and productivity.

There are at least five different kinds of nematodes that infect and cause diseases on peanut in the United States. Since there are great differences in the impact of the various nematodes on peanut quality and yield, it is very important that the nematodes in fields where peanut is to be planted be identified. Soil samples should be taken for analysis according to recommendations of the Cooperative Extension Service.

Common Nematodes

Peanut Root-Knot Nematode

The peanut root-knot nematode, *Meloidogyne arenaria* (Neal) Chitwood, is the most widespread and destructive nematode pest of peanut. There are two host races (types) of this pest, which can be distinguished only on the basis of their reaction on peanut. Race 1 infects and reproduces on peanut, whereas race 2 does not. Both races are widespread in many areas where peanut is produced, and they both cause diseases on other agronomic, vegetable, ornamental, and fruit crops. The presence of two races makes the use of routine nematode advisory services difficult because of the time required to determine the race identity.

In the United States, race 1 of the peanut root-knot nematode is common in peanut fields in Alabama, Florida, Georgia, and Texas. Sporadic occurrences have been reported in North Carolina, South Carolina, and Virginia. Numerous juveniles (infective stage) are generally distributed throughout the soil profile of infested fields. In fact, in deep, sandy soils, the largest numbers of the infective stage are generally found 12–48 inches (30–123 centimeters) deep at planting. The juveniles

are mobile and can move up from the deeper soil depths to infect plant roots, but hard pans or soils with underlying clay restrict the downward movement of the nematode. Once the juveniles infect plant roots, they mature and develop into the adult, egg-laying stage. Generally, 250–500 eggs are laid in a gelatinous matrix. Egg masses may exist in large numbers on pegs, pods, and roots during the growing season, but the number of egg masses decreases significantly after peanut harvest. Juveniles exist throughout the year and are the major survival stage.

Peanut plants in soils infested with peanut root-knot nematode generally have noticeable above- and below-ground symptoms. Definite aboveground symptoms normally appear 75–90 days after planting but may appear sooner, depending on the initial population density of nematodes in the soil and on environmental conditions. Plants that are heavily infected are stunted and yellow, and pod production is reduced (Plates 99 and 100). Root-knot disease is spotty, and plants have variable symptoms. If drought occurs near the end of the season, it may greatly increase the severity of root-knot disease, and the weakened plants may die.

To diagnose root-knot disease, the roots, pegs, and pods should be examined for the presence of nematodes. Juveniles, which are too small to see with the unaided eye, infect peanut roots soon after planting. The characteristic symptom of the disease is the abnormal swelling (galls or knots) on the roots, pegs, and pods (Plate 101). Galls and egg masses form rapidly and become apparent on the roots 55–90 days after planting. The galls on peanut roots are small and generally discrete, whereas galls on other host crops, such as tomato, squash, and tobacco, may be large and coalesced. Nematode galls can be distinguished from *Bradyrhizobium* nodules (Chapter 1), which are distinct, round swellings that appear to be attached to the root and are easily detached. Nematode galls are irregularly shaped swellings that constitute a part of the root and cannot be detached without destroying the integrity of the root. The nematode may also infect and reproduce in the *Bradyrhizobium* nodules.

Soon after blooming and the initiation of pod set (about 45

days after planting), the nematode may infect pegs and pods. Early infection may result in a weakened peg and an aborted pod, or if a pod forms, it may be detached from the plant and remain in the soil at harvest. Shells of pods that set may become heavily infected and extensively galled (Plate 102) late in the season, resulting in yield losses. Galled tissue is especially vulnerable to infection by several soilborne fungi, which greatly increases the root deterioration.

Northern Root-Knot Nematode

The northern root-knot nematode, *Meloidogyne hapla* Chitwood, is an important pest of peanut primarily in Virginia, North Carolina, Georgia, and Oklahoma. However, it may also be found as far south as southern Alabama.

> **Currently there are no peanut cultivars with resistance to nematodes.**

Although root-knot disease induced by this nematode may be substantial, it is not comparable to the devastation caused by the peanut root-knot nematode. Aboveground symptoms are less severe than those caused by the peanut root-knot nematode, and the northern root-knot nematode does not usually reduce plant growth as much as the peanut root-knot nematode. Below-ground symptoms are similar to those of the peanut root-knot nematode. However, galls induced by the northern root-knot nematode are smaller, and the root systems of infected plants are matted because of the extensive root proliferation from numerous nematode infection sites (Plate 103). Other aspects of development and etiology are similar to those of the peanut root-knot nematode.

Lesion Nematode

One of the most common nematode pests of peanut, but also the least understood and most unpredictable, is the lesion nematode, *Pratylenchus brachyurus* (Godfrey) Goodey. This nematode infects the roots, pegs, and pods, but the most conspicuous symptoms are the small, brown lesions on the pegs and pods (Plate 104). These lesions are easily confused with those induced by other soilborne pests (Chapter 11). In addition to inducing lesions on roots, pegs, and pods, heavy infection also reduces the size of the root system. Although there are few or no aboveground symptoms, the yield losses may be approximately 15–25%. The most serious problem caused by this nematode is reduced pod quality resulting from the presence of numerous lesions and from interactions with soilborne diseases. Fungi, such as *Rhizoctonia solani*, may invade the lesions and cause a rapid deterioration of the entire root system.

Sting Nematode

The sting nematode, *Belonolaimus longicaudatus* Rau, is widely distributed, occurring commonly in the sandy soils along the eastern seaboard from New Jersey to Florida and westward to Texas, Oklahoma, and Arkansas. However, most of these populations are not known to affect peanut. Populations of the nematode that parasitize peanut occur only in a limited geographical area along the Virginia-North Carolina border. Therefore, its importance in peanut production is relatively minor. However, where it does infect peanut, the sting nematode is one of the most devastating nematode pests of the peanut plant.

Sting nematode infection results in a very stunted and chlorotic plant with severely reduced root growth. The nematode feeds on the external root system, pegs, and pods and is rarely found inside the root or pod tissue.

Ring Nematode

The ring nematode, *Criconemella ornata* (Raski and Luc), is very common in many peanut fields in the southeastern United States (Florida, Alabama, and Georgia). Population densities increase rapidly, and relatively large numbers of the nematodes may be associated with peanut roots. Regardless of the population density, the nematode rarely causes noticeable aboveground symptoms, and there is little discernible loss in yield or quality.

Management

Currently there are no peanut cultivars with resistance to nematodes. Consequently, producers must rely on cultural practices, such as crop rotation, winter cover crops, and crop destruction, and the application of nematicides for management of nematodes.

Cultural Practices

Crop Rotation. Peanut should follow cultivation of crops that are not hosts or that are poor hosts of the nematodes that infect peanut. These include certain forage crops or pasture grasses. It is difficult to list specific crop-rotation schemes that are ideally suited for each of the peanut-production regions. Information on local cropping practices should be sought from the state and county peanut extension personnel. General guidelines are given here.

Planting peanut continuously or planting it after nematode-susceptible crops such as soybean, tobacco, and tomato should be avoided. There are a few corn cultivars that can reduce root-knot nematode population densities somewhat, but most cultivars are highly susceptible to lesion, sting, and ring nematodes. Some grain sorghum cultivars provide better suppression of root-knot nematode populations than corn. Cotton is a good crop to rotate with peanut because peanut and northern root-knot nematodes do not infect cotton and the cotton root-knot nematode does not infect peanut. Bahia grass is a good pasture grass to precede peanut in regions where it can be grown. Three years or more of Bahia grass that is relatively weed free should precede peanut.

The main benefit of crop rotation is that it reduces populations of nematodes, other soilborne pathogens, and insect pests. Once nematode population densities are high, effective management is very difficult. However, crop rotation schemes have some limitations. They are costly to investigate and require 1 year or longer to show benefits. In addition, they may support some nematode reproduction or reduce one nematode species while increasing another. Some weeds that are nematode hosts may even infest the rotation crop. The lesion nematode is especially difficult to control with rotation because of its wide host range.

Winter Cover Crops. Proper selection and planting of cover crops may be helpful in the management of nematodes. Cover crops can help to prevent growth of weeds that are hosts for nematodes, and they help to protect the soil from erosion during the winter. However, there are several variables that affect the benefits derived from the winter cover crops. Local populations of nematodes may have different effects on cover crops. The specific cultivar of the cover crop may also affect the nematode response. Soil temperatures during the seedling stage, throughout the winter, and during the early spring affect the potential for nematode infection and reproduction on a cover crop. If soil temperatures are relatively low after planting (below 65°F [18°C]) and remain fairly low throughout the winter and early spring, nematode infection and reproduction are suppressed. If nematodes infect the plant roots but reproduction is suppressed by low soil temperatures, then the crop is a trap crop. Rye and several oat cultivars support relatively low levels of reproduction of peanut root-knot nematodes, whereas most wheat and barley cultivars support root-knot nematodes. Local information on the nematodes and crop response to these nematodes should be obtained from state and county peanut extension specialists.

Crop Destruction. After harvest, the old plant roots should be exposed to as much sunlight and drying as possible. Otherwise, nematodes may remain active and continue to reproduce in the remaining plant roots until late in the fall. It is important to disrupt the life cycle of the nematode to prevent the production of additional infective juveniles. Each root-knot nematode female can deposit 250–500 eggs in 12–25 days, so within a short time period, one extra generation can produce billions of nematodes per acre.

The disking and shredding of plant roots initiate root decay and expose the nematodes to the numerous, naturally occurring soil antagonists that kill them. These biological control agents help to reduce the overwintering nematode population.

Chemical Control

Fumigants and Nonfumigants. One method of nematode control is the use of nematicides. Two groups of chemicals, fumigants and nonfumigants, are currently registered for nematode control on peanut. Fumigants are volatile liquids that form gases when released into the soil (Chapter 14). The fumigant penetrates soil pore spaces and moves into the water that surrounds soil particles. Nematodes in the soil moisture film are killed by contact with the fumigant.

The mode of action of nonfumigants is different. They are mostly nonvolatile, liquid or granular formulations that are carried downward with water after incorporation into the top 2–3 inches (5–7 centimeters) of soil. Nematodes are either killed on contact or their behavior is altered by the chemical, thereby affecting the parasitism.

Two types of fumigants are used on peanut. One is applied for nematode control only. The other is formulated to combat nematodes and soilborne fungal pathogens and is more costly. Several of the nonfumigant nematicides are multipurpose and suppress populations of nematodes, some soilborne insects, and piercing and sucking foliage-feeding insects (Chapter 8).

Selection of Nematicides. Growers should obtain nematicide recommendations from a local extension specialist who is familiar with the existing nematode pests. For example, the peanut root-knot nematode is a difficult pest to manage. Several of the nematicides perform poorly against this pest in deep (greater than 14 inches [35 centimeters]), sandy soil. The sting nematode is more easily controlled than the peanut root-knot nematode; thus, the choice of nematicide and the application method are not as critical.

A peanut producer should also look at the overall crop production system before choosing a nematicide because several of the chemicals have multipurpose activity. If early season control of both insects and nematodes is critical, then choosing a product that has both nematicidal and insecticidal properties is important. If the producer has experienced

difficulty in managing nematodes because of high population densities, type of nematode, or soil type, the use of soil fumigants should be considered. Because of their broad spectrum and mode of action, fumigants often provide the most effective control.

Application of Nematicides. No nematicide or other pesticide will perform satisfactorily if it is improperly applied. Dosage, timing, and placement are critical management decisions.

Fumigants must be incorporated into the soil 7–14 days before planting, either by bottom plow, deep chisel plows, or subsoiling and bedding equipment. They may be broadcast or applied in the row. In-row application at two chisels per row spaced 10 inches (25 centimeters) apart works better than one chisel per row by creating a wider treated area for pod development. In deep, sandy soils, deep placement of the fumigant and proper sealing is critical. Bottom plow application works well because the fumigant is well sealed after application. Whether they are injected into the row or broadcast, fumigants must be placed 12–14 inches (30–35 centimeters) deep, and the chisel traces must be tightly closed.

> **No nematicide or other pesticide will perform satisfactorily if it is improperly applied. Dosage, timing, and placement are critical management decisions.**

Most nonfumigants are formulated as granules. These materials must be metered uniformly, whether broadcast or in the row. The most common method of application is to place the compounds at planting in a band 7–18 inches (18–46 centimeters) wide immediately ahead of the planter shoe. If a sweep is placed just ahead of the row bander, the granules are incorporated by the roiling action of the soil from the sweep, planter shoe, and press wheel. Some nonfumigants may be applied in the seed furrow. This practice is mainly for early season insect control, although it provides some protection against nematode infection of the taproot.

Some of the nonfumigant nematicides are labeled for post-plant treatment at peg initiation. This practice affords supplemental control of nematodes in infested fields, and an increase in pod yield of approximately 15% is possible. Only granular formulations are approved for this method of application, and they should be deposited in a 14-inch (35-centimeter) band.

Selected References

Minton, N. A., and Baujard, P. 1990. Nematode parasites of peanut. Pages 285-320 in: Plant Parasitic Nematodes in Subtropical and Tropical Agriculture. M. Luc, R. A. Sikora, and J. Bridge, eds. C.A.B. International, Kew, England.

Porter, D. M., Smith, D. H., and Rodríguez-Kábana, R. 1982. Peanut plant diseases. Pages 326-410 in: Peanut Science and Technology. H. E. Pattee and C. T. Young, eds. American Peanut Research and Education Society, Yoakum, TX.

Porter, D. M., Smith, D. H., and Rodríguez-Kábana, R., eds. 1984. Introduction. Pages 1-4 in: Compendium of Peanut Diseases. American Phytopathological Society, St. Paul, MN.

David M. Wilson
Department of Plant Pathology
University of Georgia, Tifton

CHAPTER THIRTEEN

Management of Mycotoxins in Peanut

Types of Mycotoxin

"Mycotoxin" is a general term given to toxic chemical compounds produced by fungi in foodstuff. The three major genera of mycotoxin-producing fungi are *Aspergillus, Fusarium,* and *Penicillium*. Corn, peanut, and cotton are the three major crops in the United States that are affected with mycotoxin contamination. Aflatoxins are mycotoxins that are produced primarily by three species of the genus *Aspergillus*. Aflatoxins are cancer-producing chemicals (carcinogens).

Mycotoxin contamination and control in peanuts became important to the peanut industry during the 1960s. Before that time, mycotoxin contamination was generally not thought to be important. Then in 1960, more than 100,000 turkeys in England died from a disease of unknown cause, which researchers termed "turkey X-disease." Earlier, swine producers in Georgia had reported that swine that were consuming moldy peanuts or corn were being poisoned. The toxins in the moldy Brazilian peanut meal consumed by the turkeys were soon identified, and the science of mycotoxicology began in earnest.

There are four major aflatoxins, B_1, B_2, G_1, and G_2, each with a slightly different structure. Aflatoxin B_1 is the most toxic and carcinogenic. Aflatoxins M_1 and M_2 are metabolic products that also contaminate foods and feeds. The toxicity related to the peanut meal that killed the turkeys was assumed to be caused by postharvest decay. This assumption led to research on aflatoxin control related to postharvest deterioration. During the 1970s, the emphasis shifted from postharvest to preharvest studies. Current management of aflatoxin contamination begins in the field, continues throughout harvest, drying, and storage, and culminates with the manufacturing process.

During the early 1960s, losses of poultry, swine, trout, and other species attributed to aflatoxin contamination of peanut, corn, and cottonseed were reported worldwide. In 1960, the agriculture industry was not aware of the potential mycotoxin problems in peanuts. In 1990, the peanut industry in the United States established a goal to eliminate aflatoxin contamination in peanut products used for food and to minimize contamination in animal feeds. The peanut industry has for many years had a voluntary code of ethics relating to afla-

toxin, which has helped ensure a high-quality, wholesome peanut supply. The effort to eliminate aflatoxin contamination from the world's peanut supply is admirable, but success cannot be assured at present.

Management of mycotoxins in peanuts is very complex. Management of aflatoxin contamination is emphasized in this chapter because it is the economically important group of mycotoxins.

Mycotoxin-Producing Fungi

Three species of *Aspergillus* produce aflatoxins in pure culture: *A. flavus* Link:Fr., *A. parasiticus* Speare, and *A. nomius* Kurtzman, Horn, & Hesseltine. Other fungi have been reported to produce aflatoxins, but attempts to verify toxin production independently have always been unsuccessful. Several species of *Aspergillus* and *Penicillium* and a *Bipolaris* species produce the closely related metabolite sterigmatocystin, which is a precursor in aflatoxin biosynthesis.

> **Current management of aflatoxin contamination begins in the field, continues throughout harvest, drying, and storage, and culminates with the manufacturing process.**

Other mycotoxins are most often found to be produced by *Aspergillus, Penicillium, Fusarium,* and *Alternaria* species. Mycotoxins other than aflatoxin produced by *Aspergillus* species may include sterigmatocystin, cyclopiazonic acid, citrinin, ochratoxin, patulin, penicillic acid, and other less important chemicals. *Penicillium* species may produce sterigmatocystin, cyclopiazonic acid, citrinin, ochratoxin, patulin, penicillic acid, and rubratoxin. The other important genus that may produce mycotoxins in peanuts is *Fusarium*. Mycotoxins that could be formed include the trichothecenes, fumonisin, and zearalenone. However, since little research has been published on the incidence of *Fusarium* mycotoxins in peanuts, their prevalence and importance is not well established.

Mycotoxins of Concern

The aflatoxins are the primary fungal metabolites of concern to the peanut industry. Of the four major aflatoxins, B_1, B_2, G_1, and G_2, B_1 and B_2 are generally produced in peanuts by toxigenic *A. flavus* isolates, and all four are produced by toxigenic *A. parasiticus* isolates. The fungal resting bodies (sclerotia) of *A. flavus* and *A. parasiticus* may contain aflatoxins B_1, B_2, G_1, and G_2. The prevalence and importance of *A. nomius* in peanut kernels is unknown.

Aflatoxin B_1 is the most toxic and carcinogenic of the aflatoxins and is of the most concern. However, in naturally contaminated peanut lots, the incidence and ratios of aflatoxins B_1, B_2, G_1, and G_2 are variable (Table 13.1). In addition to the variable incidence of *A. flavus* and *A. parasiticus,* many of the aspergilli that are present may not be toxin producers; about 40–80% of *A. flavus* and *A. parasiticus* isolates produce toxins. For these reasons, the ratio of the G aflatoxins to the B aflatoxins in contaminated lots cannot be predicted. Because some kernels may contain B_1 and B_2 while others contain B_1, B_2, G_1, and G_2, it is important to know the total aflatoxin content. Some countries base their aflatoxin regulations on B_1 content, and others base theirs on total aflatoxin content. Many of the immunochemical and chemical screening methods do not distinguish among the aflatoxins, and others detect B_1 only, which may at times limit their usefulness.

There is an interaction between *A. flavus* and *A. parasiticus* when both occur in nature. In pure culture, *A. parasiticus* typically produces about the same amounts of B_1 and G_1, but the presence of *A. flavus* decreases the production of aflatoxins G_1 and G_2 by *A. parasiticus*. Aflatoxins G_1 and G_2 may also be less stable in contaminated peanut lots or during extraction, making it more difficult to assay for them.

> **Peanuts from randomly selected plants or rows should be collected to give a representative sample.**

Cyclopiazonic acid is another *A. flavus* metabolite that may co-occur with the aflatoxins. Cyclopiazonic acid is apparently not produced by *A. parasiticus*. It could have been present in the toxic Brazilian peanut meal associated with turkey X-dis-

ease, since peanuts naturally contaminated with cyclopiazonic acid have been reported. Presumably, this contamination accompanied aflatoxin in the turkey X-disease outbreak, both toxins having been produced by *A. flavus*. However, several other *Aspergillus* and *Penicillium* species have been demonstrated to produce cyclopiazonic acid. The toxicities of aflatoxin and cyclopiazonic acid are additive, which may be important in some animal health problems.

Rubratoxin is produced by *P. purpurogenum* O. Stoll. This fungus can invade peanuts in the field and produce rubratoxin in preharvest peanuts. However, little is known about the incidence of rubratoxin contamination of peanuts. The toxin citrinin is produced by *P. citrinum* Thom and by several other *Penicillium* and *Aspergillus* species. *P. citrinum* is frequently isolated from peanuts in the field and in storage. Citrinin has been reported to contaminate moldy peanuts in India and may occur worldwide. Citrinin also may co-occur with ochratoxin, as it does in barley.

Other mycotoxins, including ochratoxin, patulin, and penicillic acid, are produced by several *Aspergillus* and *Penicillium* species. Patulin and penicillic acid are not stable in high-protein crops and probably would not occur in peanuts, whereas ochratoxin and other mycotoxins may occur in stored peanuts. The *Fusarium* and *Alternaria* mycotoxins are not often mentioned in relation to contamination of peanuts.

Aflatoxins

Aflatoxin Analysis

Aflatoxins are currently the only mycotoxins regularly monitored in peanuts in commerce. Therefore, only aflatoxin detection and determination will be considered here. To monitor and control aflatoxin contamination effectively, it is essential to estimate accurately and conveniently the aflatoxin content. However, this is very difficult in practice. Proper sampling plans, grinding and extraction procedures, and analytical techniques all must be appropriate for each situation. Each of these problems must be approached in a logical, safety-conscious manner. Sampling of peanuts in a single field to assess preharvest contamination should also include the areas of probable plant stress, so each field should be evaluated for areas of stress before sampling.

Peanuts from randomly selected plants or rows should be collected to give a representative sample. Ideally, at least 150 pounds (69 kilograms) of peanuts should be shelled, ground to pass a 20-mesh screen, mixed, and subsampled to approximate the average aflatoxin concentration in the field. In practice, this approach is impractical and would be useful only in research situations. Sampling of peanuts in the windrow after they have been dug and inverted is easier. Areas with suspected aflatoxin contamination could be identified and

Table 13.1. Distribution and percentage of aflatoxins in naturally contaminated peanuts over several years

Aflatoxin	University of Georgia[a]		USDA[b]		FDA[c]	
	Frequency	Percent	Frequency	Percent	Frequency	Percent
B_1 only	546	43.0	0	0.0	110	30.7
$B_1 + B_2$	242	19.1	935	46.8	166	46.4
$B_1 + B_2 + G_1$	173	13.6	0	0.0	31	8.7
$B_1 + B_2 + G_1 + G_2$	176	13.9	1,063	53.1	47	13.1
$B_1 + G_1$	132	10.4	2	0.1	4	1.1
Total	1,269	100.0	2,000	100.0	358	100.0

[a] Data from contaminated, freshly harvested raw peanuts.
[b] U.S. Department of Agriculture. Data from shelled, contaminated raw peanuts from official grade samples.
[c] Food and Drug Administration. Data from contaminated, processed peanuts and peanut products collected for compliance program.

marketed separately. The usual practice for detection of afla-toxin or *A. flavus* is to sample unshelled peanuts in drying wagons after harvest and to sample commingled lots in ware-houses or shelling outruns. These samples should be taken according to established random sampling procedures.

There should be at least 25 pounds (11.5 kilograms) in a sample from a single field or drying wagon in order to ensure a representative subsample. For shelled lots, at least 150 pounds (69 kilograms) of kernels should be in a sample. Samples should always be ground and well mixed before a subsample is drawn. The subsample variability depends on the subsample size and the distribution of contamination in the lot. Averaging analytical results from several subsamples will give a more accurate assessment than a single analytical result. Various analytical techniques can be used to estimate aflatoxin content, including thin-layer chromatography (TLC), high-performance liquid chromatography (HPLC), and immuno-chemical methods. Screening methods include minicolumns,

selective adsorption of multimycotoxins (SAM), and immuno-chemical techniques.

The official methods of the Association of Official Ana-lytical Chemists (AOAC) have been rigorously evaluated in collaborative studies and should be used for peanuts in com-merce. Several assay methods, including TLC and HPLC, and immunochemical methods are AOAC official methods. Some excellent, newly developed methods have not yet been collaboratively studied by the AOAC and should be used with appropriate controls. The commercially available aflatoxin test kits are listed in Table 13.2.

Colonization by *A. flavus* and Preharvest Aflatoxin Contamination

Preharvest inoculum may reside in the soil, may be air-borne, or may be introduced on seed. The inoculum may be conidia (spores), mycelial fragments, or sclerotia, and each could serve as starting inoculum. Colonization of peanuts by

Table 13.2. Commercially available aflatoxin test kits[a]

Test kit	Analysis	Type of test	Level of detection (ng/g or ppb)	Analysis time[b] (min/sample)	Application	Contact
Aflatest	B_1, B_2, G_1, G_2	Affinity column[c]	2	7	Instrumental, quanti-tative, fluorometer, HPLC[d]	VICAM 313 Pleasant St. Watertown, MA 02172 (617) 926-7045 (800) 338-4381
AgriScreen	B_1, B_2, G_1	ELISA,[c] microtiter wells	1	12	Visual and instrumental, semiquantitative, quantitative	Neogen Corp. 620 Lesher Place Lansing, MI 48912 (517) 372-9200 (800) 234-5333
Afla-20	B_1, B_2, G_1	ELISA, cup	20	4	Visual, pass/fail	International Diagnostic System Corp. P.O. Box 799 St. Joseph, MI 49085 (616) 428-8400
Afla-10	B_1, B_2, G_1	ELISA, cup	10	4	Visual, pass/fail	
CITE-Probe-Aflatoxin	B_1	ELISA, probe	20	3	Visual, pass/fail	IDEXX 100 Fore Street Portland, ME 04101 (207) 774-4384 (800) 548-6733
EZ-SCREEN: aflatoxin	B_1, B_2, G_1	ELISA, card	20	7	Visual, pass/fail	EDITEK, Inc. P.O. Box 908 1238 Anthony Road Burlington, NC 27215 (919) 226-6311 (800) 334-1116
SAM-A	B_1, B_2, G_1, G_2	Selective adsorption[e]	10	10	Visual, pass/fail	Dr. Timothy D. Phillips College of Veterinary Medicine Texas A&M University College Station, TX 77843-4458 (409) 845-6414
SAM-AZ	B_1, B_2, G_1, G_2 (zearalenone)	Selective adsorption	10 500	10	Visual, pass/fail	
HV Minicolumn	B_1, B_2, G_1, G_2	Minicolumn	20	10	Visual, pass/fail	Romer Labs, Inc. 1301 Stylemaster Dr. Union, MO 63084 (314) 583-8600 (800) 769-1380

[a] Adapted from Anonymous, 1989.
[b] Does not include sample preparation and extraction.
[c] Immunochemical methods: affinity column or ELISA (enzyme-linked immunosorbent assay).
[d] High-performance liquid chromatography.
[e] Modified Holaday-Velasco minicolumn (AOAC method 975.36).

toxigenic strains of *A. flavus* or *A. parasiticus* is necessary for subsequent aflatoxin contamination.

Virgin soils typically do not contain propagules (fungal structures) of *A. flavus* or *A. parasiticus*, while cultivated soils frequently do contain such viable propagules. Airborne inoculum could also contribute to colonization of flowers. Infection or colonization by *A. flavus* or *A. parasiticus* could involve active fungal growth, or the colonized tissue may contain fungi that are not actively growing. Insects such as termites or the lesser cornstalk borer possibly carry inoculum to the peanut tissue and also create favorable habitats for fungal growth.

Colonization of peanuts by toxigenic strains of *A. flavus* or *A. parasiticus* is necessary for subsequent aflatoxin contamination.

Preharvest colonization of peanut kernels by *A. flavus* and *A. parasiticus* is usually less than 2% in undamaged pods during years with adequate rainfall and moderate temperatures. Pods mechanically damaged or damaged by insects or disease are more frequently invaded by *A. flavus*. Drought stress during the last 30–45 days of the growing season increases the incidence of *A. flavus* in both damaged and undamaged pods. If the soil temperature is 76–90°F (25–32°C), aflatoxin may be produced in undamaged, food-quality peanuts under drought conditions. Adequate soil moisture at these temperatures allows fungal colonization but no aflatoxin accumulation. Moisture from rainfall or irrigation usually lowers soil temperatures in the field, which may partially explain the soil temperature and moisture interaction. However, moisture may also influence the plant's response to infection. For example, there is evidence that hydrated peanuts produce phytoalexins, which are chemicals produced in response to fungal invasion that can be toxic to or inhibit the growth of fungal pathogens. *A. flavus* and *A. parasiticus* can invade damaged peanuts in the ground and produce aflatoxin in damaged pods regardless of the soil moisture or temperature.

Pods and kernels damaged by soil-inhabiting insects, including the lesser cornstalk borer and termites (Chapter 8), can greatly enhance aflatoxin contamination. Pod damage, without penetration, by the lesser cornstalk borer increases invasion of kernels by *A. flavus* but does not always increase aflatoxin content. Other soil-inhabiting insects, especially those active in dry soils, can also affect aflatoxin contamination by either acting as vectors of *A. flavus* or creating a favorable habitat for fungal growth. Damage by pod-rotting fungi sometimes also increases aflatoxin contamination by creating a favorable habitat for *A. flavus*.

Although no data are available on the effects of plant stress on aflatoxin in peanut, that caused by high or low plant densities, disease, weeds, and overfertilization could affect the incidence of aflatoxin contamination. These factors influence contamination of corn, and similar effects might be seen in peanuts. Mechanical damage of pods by farm implements also increases the probability of aflatoxin contamination.

Postharvest Aflatoxin Contamination

Once peanuts have been dug, the environment surrounding the pods is completely changed. In the United States, peanuts are dug and placed in inverted windrows to facilitate drying. There is ample opportunity for fungal invasion to occur in the windrows, but in practice, little additional colonization by *A. flavus* or *A. parasiticus* takes place (Plates 105 and 106). During 4 years of study in Georgia, no increases in *A. flavus* incidence or aflatoxin contamination were noted in inverted peanuts, and no findings of increases have been published by other researchers. Therefore, wet conditions leading to invasion would have to be very prolonged to increase aflatoxin contamination. These prolonged wet conditions would probably decrease overall quality rather than substantially increase aflatoxin content. In contrast, research published on stacked peanuts and other slow, field-drying methods indicates that these drying methods allow *A. flavus* growth. Aflatoxin contamination can result from slow drying methods or when artificial drying in wagons is ineffective or not used.

After peanuts have been dug, the moisture content must be reduced for marketing and safe storage (Chapter 5). Drying is usually done first in the windrow to 20–25% moisture on a fresh weight basis followed by combining and artificial drying to an in-shell moisture content of 10–11%. Aflatoxin production can be very rapid in drying wagons if drying is slow because of poor air movement or low temperatures. Pods, roots, plant parts, and foreign material with high moisture contents can create a favorable habitat for *A. flavus* growth and localized areas with high amounts of aflatoxin contamination (Plate 107). Loose-shelled kernels and damaged kernels and pods are the most likely to be contaminated during drying.

Aflatoxin contamination can result from slow drying methods or when artificial drying in wagons is ineffective or not used.

Moisture content is the single most important variable in stored peanuts. *A. flavus* will grow and produce aflatoxins in stored peanuts if the moisture content is sufficient. Workers at Auburn University determined that the critical maximum relative humidity for safe storage is 84 ± 1% at 86°F (30°C). This roughly corresponds to a moisture content of 10–11% in unshelled peanuts and 6–8% in shelled peanuts. *A. flavus* and *A. parasiticus* will not grow under these conditions, but other fungi can grow slowly at lower levels of humidity. It should be noted that hot spots related to high-moisture fragments, insect activity, and microbial growth can develop even under these conditions. The moisture content of peanuts in storage is affected by temperature, relative humidity between kernels, insect and rodent activity, microbial growth, water leaks, and moisture migration. All of these factors must be considered when unshelled or shelled peanut stock is warehoused.

Management of Aflatoxin Contamination

Preharvest Strategies. Management strategies for minimizing or eliminating aflatoxin contamination begin in the field and end with the manufacturer. The aflatoxin problem can be described as a preharvest problem, a harvesting and drying problem, and a warehousing and utilization problem. Each of these phases is a separate problem, and each provides a different environment for creating a favorable habitat for fungal growth. Management of preharvest contamination at present must rely on production practices (Box 13.1).

Fungicides, including the sterol-inhibitor compounds (Chapter 11), have little or no measurable effect on aflatoxin contamination when they are used to control plant diseases. Fungicides sprayed on inverted peanuts in the windrow have little beneficial effect. The costs of such windrow applications are prohibitive, and chemical residues could be problematic. However, rotations, proper plant densities, and general control of pests, including nematodes, weeds, plant disease, and soil-inhabiting insects, may improve plant vigor and thus help minimize the risk of aflatoxin contamination.

Soil-inhabiting insects that damage pods, such as the lesser cornstalk borer in the United States and the termite in Africa, can increase aflatoxin contamination of damaged kernels and infection of undamaged kernels by *A. flavus*. Therefore, every precaution should be taken to control soil-inhabiting insects that are active during pod maturation. These insects may be more active under drought conditions. The effects of planting date on insect activity and aflatoxin contamination need to be defined to see whether planting date is related to contamination.

Adequate mineral nutrition is important for maximum peanut production, and the factors that increase quality might also reduce contamination. Calcium is the only nutrient that has been shown to have an effect on aflatoxin contamination. In studies at Tifton and Dawson, Georgia, calcium deficiencies sometimes increased contamination in harvested peanuts. Application of calcium could be less effective under drought conditions because the lack of water inhibits calcium uptake by the pod. *A. flavus* infection is more frequent in calcium-deficient soils, but this infection does not always result in contamination. For these reasons, recommended rates of calcium should be used. However, applications at rates higher than recommended do not further decrease *A. flavus* colonization or aflatoxin content. Biological control of *A. flavus* by the fungus *Trichoderma harzianum* Rifai in research plots has been shown to be slightly more effective in soils with sufficient calcium than in calcium-deficient soils. More research is necessary on biological control before conclusions can be drawn about its potential use.

Late-season drought stress with soil temperatures of 76–90°F (25–32°C) is by far the most important single determinant of preharvest aflatoxin contamination. The critical drought period occurs during the last 45 days of the growing season. Longer drought stresses increase aflatoxin contamination and also dramatically decrease yield. Drought stresses earlier in the growing season affect yield, maturity, and quality but do not seem to have much effect on aflatoxin production unless the drought lasts until harvest. Moisture during the last 30 days causes the contaminated pods to deteriorate and remain in the soil at harvest. In order to control preharvest contamination, it is essential to avoid an extended drought during the last 45 days of the season. A 20-day drought stress late in the season was not seen to enhance aflatoxin contamination in studies at the U.S. Department of Agriculture National Peanut Research Laboratory. Even if a routine practice of irrigation 25 days before harvest preceded by a drought were carefully followed, aflatoxin could still be present in immature and damaged pods and kernels.

Postharvest Strategies. After peanuts have been dug, measures to control aflatoxin depend mainly on moisture management. Windrow drying is weather dependent, but little additional aflatoxin occurs in inverted windrows unless they are wet for an extended period. Once the peanuts have been combined, aflatoxin content can rapidly increase if they are kept above 11% moisture in the drying wagon. Rapid and uniform drying in wagons is crucial for aflatoxin management. Removal of immature pods, loose-shelled kernels, roots, weeds, and foreign material during harvest is important, not only because these components are frequently contaminated but also because they may be difficult to dry uniformly.

Peanuts may be stored safely in the shell if moisture is below 10–11%. Moisture can be increased in storage by moisture migration, water leaks, and biological activity. Therefore, sanitation, aeration, and maintenance are all essential for aflatoxin management. Prevention of aflatoxin accumulation in storage is related to maintenance of proper moisture content in the stored product. Peanuts may be stored in atmospheres with low oxygen or high carbon dioxide levels to control *A. flavus* growth and insects.

Removal of Aflatoxin Contamination

Adjustments to the peanut combine are the first variables at harvest that can be used to physically remove aflatoxin-contaminated peanuts. Damaged pods, loose-shelled kernels, and immature peanuts are the most likely to be contaminated, so the combine should be adjusted to minimize damage and remove these sources of contamination as well as foreign material. During years in which aflatoxin contamination is associated with drought, sound, mature kernels are also likely to be contaminated.

The use of the belt-screening technique to remove foreign material, small, immature pods, and loose-shelled kernels could be of great benefit in aflatoxin removal. During the Peanut Quality Enhancement Project in 1988, Spanish peanuts

---

Box 13.1

Factors that Affect Aflatoxin Contamination of Peanuts

Preharvest
- Drought and irrigation
- Calcium availability
- Soil-inhabiting insects
- Soil temperature
- Biological damage
- Mechanical damage
- Cultivars (undefined effects)
- Nematicides (no effects)
- Fungicides (few effects)

Postharvest
- Artificial drying
- Moisture content
- Water leaks
- Moisture migration
- Temperature
- Insect pests
- Rodents
- Microbial deterioration

were cleaned by using a 22/64-inch screen opening, and runner and Virginia peanuts were cleaned with a 24/64- or 26/64-inch screen opening. Analyses of data on the aflatoxin content of these peanuts demonstrated that screened peanuts contained less aflatoxin overall than unscreened peanuts. Removal of oil stock and damaged kernels resulted in further reductions in aflatoxin. The belt screen was less effective on peanuts from a severely drought-stricken area than on peanuts grown under adequate rainfall because more mature kernels were contaminated.

Currently, the visual inspection method for detecting *A. flavus* conidial heads is used to segregate peanut lots suspected of containing aflatoxin at the buying point. This is not a chemical test, and many uncontaminated lots are diverted into the Segregation III oil stock category, and more importantly, many contaminated lots may not be identified. A direct chemical test at the buying point would be fairer and more accurate. Immunochemical techniques are available that make this possible, and these techniques will certainly be improved. A direct aflatoxin detection method would also improve the ability to warehouse suspect lots separately. The same sampling and subsampling problems discussed earlier make implementation of such tests dependent on improvements and changes in the grading system.

The shelling process includes several steps that help remove aflatoxin contamination from Segregation I peanuts. During the precleaning operations, loose-shelled kernels and immature pods, which are diverted to nonfood uses, and foreign materials are removed. After precleaning, the peanuts are shelled and sized or cleaned further if they are to be sold as in-shell peanuts. Sized peanuts are separated into the various edible and oil stock grades. Edible peanuts are generally sorted by hand on a picking table, with electric color sorters, or by a combination of the two. The pickouts are diverted to oil stock. Careful hand sorting is more effective than color sorting, so a combination is preferred.

Presently in the United States, raw peanut lots are sampled according to a defined protocol and analyzed after shelling according to an AOAC official method that is also accepted by the U.S. Department of Agriculture. If the aflatoxin content is below 16 nanograms per gram (parts per billion [ppb]), the lot is allowed into commerce. If the lot contains between 16 and 75 ppb total aflatoxin, more samples are analyzed as defined by the protocol; if the average is 25 ppb or less, then the lot is allowed into commerce. Lots containing more than 25 ppb on average can be reprocessed for further aflatoxin reduction. Recleaning and sorting raw peanut lots sometimes removes enough of the contamination to allow the lot into commerce. If recleaning is not successful, the peanuts can be blanched.

Blanching to remove aflatoxin contamination is a specialized process during which the peanuts undergo a white roast and the skins are removed. The blanched peanuts are then color sorted. Almost always, the peanut lots that are formed after the final color sort contain very small amounts of aflatoxin because peanuts containing any mold or damage darken during the roasting process, making the final color sort very efficient. Blanching for aflatoxin removal sometimes results in as much as 25% shrinkage, but the residual peanuts are usually free of aflatoxin.

Roasting of peanuts by manufacturers also helps remove some of the contamination because the high roasting temperatures may destroy up to 50% of the aflatoxin. Ammoniation is the preferred way to detoxify peanut meal for use as animal feed. However, the Food and Drug Administration has not approved ammoniation for use in the United States.

In summary, aflatoxins are the mycotoxins of most concern because they are toxic and carcinogenic. Aflatoxin regulations regarding peanut products are not uniform worldwide, and management strategies must take this and the variances associated with sampling and analysis into account. Preharvest aflatoxin contamination is affected by the fungal strain, inoculum density, soil moisture, soil temperature, calcium availability, and biological or mechanical damage. Contamination after the peanuts have been dug depends primarily on moisture content. Peanuts should be dried and stored at 10–11% in-shell moisture content. Storage structures should be clean, weather tight, and free of insect and rodent infestations. Damaged peanuts can be removed by sorting during the shelling or blanching process. These separation methods physically remove the majority of contaminated kernels and improve the quality of the finished product.

Selected References

Anonymous. 1989. Mycotoxins: Economic and health risks. Counc. Agric. Sci. Technol. Rep. 116.

Dorner, J. W., Cole, R. J., Sanders, T. H., and Blankenship, P. D. 1989. Interrelationship of kernel water activity, soil temperature, maturity, and phytoalexin production in preharvest aflatoxin contamination of drought-stressed peanuts. Mycopathologia 105:117-128.

Goldblatt, L. A., ed. 1969. Aflatoxin—Scientific Background, Control, and Implications. Academic Press, New York.

Helrich, K., ed. 1990. Official Methods of Analysis, 15th ed. Association of Official Analytical Chemists, Arlington, VA.

McDonald, D., and Mehan, V. K., eds. 1989. Aflatoxin Contamination of Groundnuts. International Crops Research Institute for the Semi-Arid Tropics, Patancheru, India.

Pattee, H. E., and Young, C. T., eds. 1982. Peanut Science and Technology. American Peanut Research and Education Society, Yoakum, TX.

Rodricks, J. V., Hesseltine, C. W., and Mehlman, M. A., eds. 1977. Mycotoxins in Human and Animal Health. Pathotox, Park Forest South, IL.

Wicklow, D. T., and Donahue, J. E. 1984. Sporogenic germination of sclerotia in *Aspergillus flavus* and *Aspergillus parasiticus*. Trans. Br. Mycol. Soc. 82:621-624.

Thomas A. Kucharek
Department of Plant Pathology
University of Florida, Gainesville

CHAPTER FOURTEEN

Pesticide Application Techniques for Peanut Health Management

Management of peanut pests requires the integration of cultural, biological, and chemical control practices. Careful application of pesticides is often necessary to manage diseases, insects, and weeds. Proper pesticide application techniques maximize pest control, minimize damage to the crop and adverse effects on the environment, prevent legal confrontations, and reduce overall production costs. Also, if correct application techniques are used and the level of control for some pest is less than expected, the user can more confidently examine other areas of crop production for the source of a problem.

The pesticide application techniques discussed here are limited to the logistics of delivering the pesticide from equipment to the soil or to the peanut crop. If the right material is used at the proper time and rate and with the correct technique, effective control can be achieved. If all of these conditions are satisfied and expected control is not achieved, it must be determined whether the target pest has been correctly identified or whether changes in sensitivity by the pest to the chemical have occurred. Even if a pesticide label claims that a certain pest can be controlled by that compound, the level of control may not be as high as it could be with a different material. Also, some pesticides, particularly those that are applied to the soil, perform differently in different situations. The local extension service should have pertinent information on the relative performance of different pesticides against plant pathogens, insects, and weeds for a given geographical area. Their information takes into account local soil conditions, weather, cultivars, and other crop-production techniques.

> **Anyone who handles pesticides should wear proper clothing for protection.**

Effective pest management is influenced by pesticide application technique, crop-production practices, and environmental variables. As the producer becomes more effective in manipulating these variables into a usable system, the level of pest control will increase. Eradication of pests is neither attainable nor always desirable because it would be costly and impractical and could be hazardous to the environment.

Details on safe use and storage of pesticides are too lengthy and cannot be presented here, but such information is available on product labels. This information should not be ignored. Anyone who handles pesticides should wear proper clothing for protection. Gloves, goggles, boots, a long-sleeved shirt, and a hat are minimal attire, and other gear may be required as indicated by the product label. Protective coveralls are readily obtainable and are essential for many pesticide operations. During mixing and loading, the individual should also wear an open-sided face shield and load with the wind at his back (upwind). Protective equipment must be clean, well maintained, and readily available according to the worker protection standard set forth by the U.S. Environmental Protection Agency. The prescribed equipment must also be worn by workers entering an area where a pesticide that requires a "restricted entry interval" has been applied.

Pesticide Formulations

Pesticides can be formulated as gases, liquids, or solids. Both liquid and solid formulations are used in peanut production. Sometimes pressurized liquids (e.g., methyl bromide) that become gaseous upon release to the air are used for fumigation of storage structures for peanuts. Liquid and solid pesticides used in field production will be discussed here.

Liquid Soil Fumigants

Fumigants are used to reduce pests that reside in the soil. A fumigant is usually injected into the soil, where it volatilizes into a gas that dissipates into the air spaces within the soil. Liquid fumigants include multipurpose chemicals (e.g., vapam) that reduce the population of nematodes, insects, fungi, bacteria, and weed seeds that exist in the soil. Liquid fumigants also include those that are generally considered nematicides (e.g., 1,3-dichloropropene [1,3-D]). Such products may have some direct or indirect effects on insects and fungi.

Dusts (D)

Dusts are applied in a dry form without dilution in water. Before the 1970s, they were used commonly for seed treatments and foliar sprays. Dusts were cumbersome to use because they commonly drifted with the wind, and a large volume of material was required (e.g., 20 pounds per acre). Dusts contained only 3–10% active ingredient and were less effective than sprayable formulations.

Wettable Powders (WP)

Wettable powders are manufactured as dry powders and are applied to the crop after dilution in water. The resulting suspension has to be constantly agitated in the spray tank to prevent settling to the bottom. Wettable powders contain 25–83% active ingredient, allowing the user to handle less weight of actual material, and are more effective than dusts because they adhere to the foliage longer.

Soluble Powders (SP)

Soluble powders are similar to wettable powders except that the chemical becomes a solution rather than a suspension when added to water. This reduces the requirement for agitation while it is in the spray tank because the material is not likely to settle out with normal use patterns. However, when a SP is tank mixed with a WP, tank agitation is required.

Flowables (F or FL)

Flowables are liquids that contain the dry fungicide, special suspension chemicals, and water. Generally, flowables do not require as much agitation in the spray tank as wettable powders. Many pesticides are available as wettable powder and flowable formulations. Liquid types are less likely to drift during loading operations and are more useful in the closed pesticide transfer systems that are becoming popular.

Prills

Prills are solid beads that become suspensions when placed in water and therefore require agitation in the spray tank. Prills are also called dispersible granules (DG), dry flowables (DF), or wettable dispersible granules (WDG). The amount of active ingredient can be 25–90%.

Emulsifiable Concentrates (EC)

Emulsifiable concentrates are liquids that contain the active ingredient, an emulsifying agent, and a solvent (e.g., xylene). When an EC is added to water, an emulsion is formed in which the active ingredient is in small oil globules suspended in the water. Minimum agitation is required to maintain an emulsion. The amount of active ingredient in these formulations generally is 1.5–8 pounds per gallon of product. Because of the oily emulsifying agent and the solvents in EC formulations, phytotoxicity to peanut leaves is more likely to occur with an EC than with other formulation types.

Granules (G)

Granular formulations are beads that contain the active ingredients in a clay, gypsum, or organic (e.g., corn cob) base. The amount of active ingredient for different pesticides varies from 2 to 20 pounds per 100 pounds of formulation. These products are applied in a dry form with specially designed granular applicators.

Pesticide Application

Amount of Pesticide to Use

Although some labels are written more clearly than others, one intent of the label is to instruct the user about the amount (dose) of material to be applied to a given area. Area is measured in acres or hectares (metric), and labeled rates are the legal rates. Rates in excess of labeled rates are more likely to cause adverse effects to the crop, the applicator, and the environment. As is often said, the label is the law.

Pesticides that are to be applied to the soil usually have a label that also gives the rate for a broadcast application, which includes the entire land area where the crop will be grown and the space between rows. The space occupied by alleys and ditches should not be included when areas for pesticide applications are calculated. Thus, if you intend to spray a preplant herbicide to the soil on a broadcast basis, the labeled rate for the broadcast application will relate to the 43,560 square feet per acre that will be occupied by the crop and space between the rows.

The label is the law.

Labels of some soil-applied pesticides have instructions regarding how to treat a proportionately smaller area by banding the application. For a banded application, the rate of the chemical applied within the treated area is the same as that of a broadcast application, but the proportionately lower amount that is used relates to the smaller area within the crop area actually being treated. For example, suppose a label instructs the use of a 12-inch band along row centers that are planted 36 inches apart. If a broadcast rate of 3 pounds per treated acre is needed, then 1 pound of material should be applied in that geographical acre. The key words are "treated acre." Labels of soil-applied herbicides, nematicides, insecticides, and fungicides contain such instructions. The following formula provides the mathematical relationship: (band width in inches/row spacing in inches) × broadcast rate per acre = amount needed per acre of field in the band. For example,

(12 inches/36 inches) × 3 pounds per acre = 1 pound per acre.

Often, the labels of granular materials conveniently indicate the rates per 1,000 linear feet of row. Where this information is provided, the above formula is not needed. Also, well-written labels will provide rates per 1,000 linear feet for different tractor speeds and for different types of applicators (e.g., Gandy, John Deere, and Noble) along with some suggested gauge settings. Calibration of equipment should be performed to confirm that the suggested gauge setting is correct.

For soil-injected fumigants, the principle is the same when proportionately less material is used for in-row treatments, but the labels may provide a specific formula or table that takes into account fumigant movement in the soil from the injection point for different row spacings. When a liquid fumigant is applied to a limited area along the row, it is actually being banded on the basis of broadcast equivalent rates. Fluid rates for each 1,000 linear feet of row are on the label. Therefore, proportional reduction of rates from a broadcast acre, as calculated for granular applications, can be done with the formula above for banded rates of fumigants. This is mentioned because some labels of fumigant nematicides may not take into account row spacing when rates per 1,000 linear feet are included.

Labels of insecticides and fungicides for foliar sprays usually indicate rates per acre but rarely use terms such as "treated acre" or "broadcast acre." The assumption is that a broadcast acre is implied because many insecticides and fungicides used for foliar pest control are delivered from aircraft, broadcast boom sprayers, air-blast sprayers, or irrigation systems. With aircraft, air-blast sprayers, and overhead chemigated applications, the rate per acre must be a broadcast rate because directed or banded sprays cannot be achieved with these types of equipment.

Spraying the soil for foliar pests is wasteful.

Unfortunately, the rate per treated acre for directed foliar sprays according to label instructions may be subject to interpretation. Does the rate on the label pertain to the entire geographical area, including the area between the rows, or does the label imply that the rates should be proportionally reduced on the basis of actual area covered by the plant canopy? Some growers spray the entire field area with a broadcast boom arrangement from the beginning of the season to the end. Spraying the soil for foliar pests is wasteful. The grower can use a proportionately lower rate per geographical acre and spray only the area currently occupied by the crop canopy. Conversely, the broadcast rate may be applied in a directed spray to the crop canopy. Although many labels do not contain a statement about this dilemma, the use of the broadcast rate for a directed spray is implied when labels provide information about nozzle arrangements for directed sprays while indicating that the higher rates of the chemical are used for mitigating higher pest severities. In Table 14.1, the level of leaf spot control is presented from two experiments in which full labeled rates of chlorothalonil were compared with proportioned rates based on the percentage of the soil covered by the peanut canopy. Because an effective fungicide was used, the level of leaf spot control exceeded 93% in all treatments. The sprays were directed at the existing canopies by the use of swivel nozzles. In this case, the interpretation of label dosage did not affect the level of control, but with other chemicals or situations, particularly when herbicides are used, label interpretation may be critical.

Critical Zone Concept for Pesticide Application

Most pests infect (e.g., fungi and nematodes) or infest (e.g., insects) certain parts of the peanut plant. The predominant locations of uncontrolled weeds depend upon field history and crop competition (Chapter 7). When pests tend to predominate

or aggregate in certain locations, the grower has an opportunity to reduce pesticide usage or attain more effective control if pesticide applications are directed to those critical zones.

Seed treatments must be used for peanuts; otherwise, poor stands will result. The critical zone for seed treatment is the location of the seed in the furrow and the soil surrounding the seed during germination and seedling emergence. Seed-treatment fungicides that are not systemic, such as captan, provide some control for the seed and preemergence stages of the seedling against several fungal pathogens that reside in the soil. A systemic chemical such as carboxin provides some protection for the young seedling, even after emergence. Usually, systemic seed-treatment chemicals provide protection for a short time after emergence, but the spectrum of fungi controlled is less than that of a broad-spectrum fungicide.

Peanut leaf spot diseases and southern stem rot (white mold) occur primarily along the row center where the peanut canopy is most dense. The lesser cornstalk borer is another problem along the row center within the pegging zone. Peanut nematodes begin their life cycle in roots along the row center early in the development of the crop and spread to pods and other roots later. On the other hand, weeds tend to be more numerous between the rows, where crop competition is delayed. However, weeds can be numerous along row centers between plants, particularly if the seeding rate or seedling emergence is low.

Protecting vine growth between the rows is important because vines are a major source of food for the pods. Likewise, peanut roots exist between the rows and cannot be ignored. However, the origin of a peanut crop is along the row center, and because many pests begin or increase in number near the row center, that zone should be the target of significant attention. Although all parts of the peanut plant contribute to the final yield, the canopy zone along the row center and the pegging zone are the critical zones for pesticide applications.

Seed Treatment

Most seed treatments for peanut are applied by the seed producer or a designated seed handler. Seed-treatment chemicals are formulated as wettable powders, flowables, or dusts. Dusts used for seed treatments are more hazardous to the individuals who treat and bag the seed than wettable powders or flowables, which can be used as wet slurries by adding small amounts of water to the formulations. Seed are conveyed through a closed slurry treater where sprays of the wet seed treatment are uniformly applied. Slurry treatments adhere to

Table 14.1. Control of leaf spot diseases on Florunner peanut in Florida with chlorothalonil at the full labeled rate and at a rate directly proportional to the full rate determined by plant size[a]

| Treatment[b] | 1982 | | | 1983 | | | |
| | Number of leaf spots | Leaflets absent | | Number of leaf spots | | Leaflets absent | |
	75 days	95 days	107 days	109 days	137 days	123 days	137 days
Unsprayed	290	27	35	374	...	18	36
Full rate	4	1	<1	16	13	0	<1
Proportional rate	3	<1	1	12	10	<1	<1

[a] Diseases controlled were early leaf spot, caused by *Cercospora arachidicola*, and late leaf spot, caused by *Cercosporidium personatum*. Assessments consisted of numbers of leaf spots and/or leaflets defoliated from 10 leaves (each leaf has four leaflets).

[b] Full rate was 2.125 pints per acre (pt/A) of chlorothalonil applied at 30 pounds per square inch at 14-day intervals with three D2-25 hollow-cone nozzles per row. Proportional rates based on percentage of soil surface covered by plant growth were 1.063 pt/A (application 1), 1.771 pt/A (application 2), and 2.125 pt/A (applications 3–5) in 1982. In 1983, the proportional rates were 0.553 pt/A (application 1),1.339 pt/A (application 2), 1.764 pt/A (application 3), and 2.125 pt/A (last four applications).

the seed coat better than dusts because sticking agents are incorporated within the wettable powder and flowable formulations. Sticking agents also reduce the amount of dust formed during handling. However, problems are sometimes encountered during plant emergence when seed have been coated with a slurry treatment. Therefore, dust formulations are still applied to some of the peanut seed used today.

Regardless of the seed treatment used, it is imperative that the person loading the seed and pesticide wear protective clothing and equipment. Individuals are most likely to come into direct contact with pesticides during the mixing and loading operations. Treated seed is classified as a pesticide by law. Dyes (e.g., red, green, and blue) are added to the seed treatment as warning agents, and the dye must be present if the seed are to be transported in interstate commerce.

Soil Fumigation

Soil fumigants are used for peanut prior to planting to control nematodes, soilborne pathogens, and soilborne insects. Because of environmental issues with dibromochloropropane (DBCP) and the contamination of groundwater with water-soluble liquid nematicides such as ethylene dibromide (EDB), these two pesticides, the most effective liquid fumigant nematicides, are no longer registered for use with peanut or other crops. Because nematodes reside and migrate in water in the soil, an effective nematicide must be water soluble to attain adequate control. The currently available liquid soil fumigant is metam-sodium (e.g., Vapam). Multipurpose fumigants reduce nematodes, soil fungi, soil insects, and weeds. The last remaining liquid soil fumigant used specifically for nematodes contains 1,3-D (e.g., Telone products). This material also has some insecticidal effects.

Applications of the currently available soil fumigants must be made at least 1–2 weeks before planting. Fumigants volatilize within the air spaces in the soil and are retained for some time by the soil and the water in the soil. Therefore, soil should have a loose tilth to allow vapors to move from the injection point. Fumigant movement within soil is restricted by organic matter and clay. Excessively warm soils promote rapid vapor movement. If the vapor moves too rapidly and dissipates from the soil, nematodes may not be exposed sufficiently to be killed. Conversely, cold soils reduce vapor movement, resulting in poor control. If the soil is not cold or excessively wet when the material is injected, 2 weeks should be adequate for the material to reduce nematode populations and dissipate from the soil. Currently available soil fumigants will be injurious to the seed or seedlings if the waiting period is inadequate.

> **The preplant injection of a liquid nematicide is important if nematodes are to be controlled because 75% of the peanut root system is formed within 7 weeks of planting.**

Injection of soil fumigants can be made on a broadcast basis, but such a treatment is too expensive and unnecessary for proper control. Soil fumigants move 6–8 inches in the soil from the injection point, depending upon soil type, soil moisture, and characteristics of the chemical. If the entire broadcast equivalent rate for two chisels is applied in one chisel along the row center, the fumigated zone will extend slightly beyond the expected 12-inch fumigation zone for one chisel (6 inches on each side). However, the most effective way to fumigate the entire pegging zone (e.g., 24 inches) is to straddle two chisels along the intended row center and use the broadcast equivalent rate for the area actually treated. The preplant injection of a liquid nematicide is important if nematodes are to be controlled because 75% of the peanut root system is formed within 7 weeks of planting. Injecting the fumigant at the base of a bottom plow has proved effective, particularly in very sandy soils.

Broad-spectrum fumigants such as metam-sodium have been used on a limited basis for peanut production. Currently, they are used in the Virginia-Carolina peanut area for control of Cylindrocladium black rot. The application techniques and soil conditions that are best for liquid nematicides are required for these fumigants as well.

Guidelines for soil fumigation are summarized in Box 14.1, and a fumigation rig is shown in Plate 108.

Granular Pesticides

Granular pesticides are applied by special applicators that usually consist of a hopper box with a fluted or ribbed rotor axle at the bottom through which the pesticide is gravity fed (Plate 109). The source of the torque to turn the rotor is either a ground-driven wheel or an electric motor. The application rate is affected by the size, shape, and density of the granules, the ground speed, the size of the adjustable gate openings, and the rotor speed. Thus, for a granular applicator that is ground driven and has a metering cam (e.g., Gandy), tractor speed and the size of the gate opening strongly influence the flow rate.

Box 14.1

Guidelines for Soil Fumigation

Soil fumigants commonly used for peanut can be delivered by gravity or via positive displacement from a pump through a manifold with disk orifices to regulate the flow rate. Smaller disk orifices reduce flow rate, and larger disk orifices increase flow rate. The metering disks, tractor speed, and pump pressure regulate the amount of chemical injected into the soil. From the manifold, the liquid moves through tubes (e.g., of polypropylene) that are inserted behind the chisel that opens the injection furrow in the soil (Plate 108). Coulters in front of the chisel reduce the amount of trash that accumulates on the chisels and final seedbed surface. The furrow is then covered with soil by bedding, coulters, or a drag to reduce rapid loss of vapors from the soil. The sealing method for coarse, sandy soils probably should include some compaction to reduce rapid vapor loss. However, in fine soils that have more clay content, compaction might lengthen the interval between injection and planting. An injection depth of 6–9 inches may be sufficient on fine soils, but a depth of 12 inches may be necessary on sandy soils since shank furrows do not seal as readily on the coarse soils.

The operator of a granular applicator should determine the variables that influence flow rate for his own granular applicator, since many designs are on the market.

It is a common misconception that ground speed does not affect application rates with granular applicators. As long as the flow rate through a device is constant at a gate opening setting or rotor speed, higher speeds proportionately reduce application rates to the treated area. Excessive rotor speeds can grind the granules to powder, affecting the application rate. When a large gate opening is used, the rotor speed in the hopper box must be fast enough to supply the increased gravitational demand. Different rotor designs affect the magnitude of influence from application speed.

> **It is a common misconception that ground speed does not affect application rates with granular applicators.**

After the granules fall through tubes from the hopper box and gate openings, they are directed to strategically placed row banders for soil-surface applications or to tubes for in-furrow applications. Row banders of various widths are available. Baffles inside the banders distribute the granules across the entire width of the row bander. When the granules fall through the peanut canopy, their distribution will be somewhat wider than the row bander because they will bounce off the leaves. If just a tube is used with no row bander, a band approximately 4 inches wide at the base of the plants will occur.

All granular treatments are directed to the soil. Therefore, when postemergence pesticides are applied, the foliage should be dry because granular pesticides will stick to wet foliage. Granules should not be incorporated into the soil unless the label and local recommendations direct such an operation. This is particularly true with postemergence applications because mechanical incorporation may damage the root system.

Granular applications are used in peanut production primarily to reduce soilborne insects, nematodes, and fungal plant pathogens. Because most currently available sprayable pesticides do not move downward in the peanut plant and low- or high-pressure sprayers are not capable of placing an adequate amount of pesticide on the soil through a peanut canopy, granules provide a method by which the entire application can be placed on the soil surface through dry foliage. This has become particularly useful for mid- to late-season control of stem rot (*Sclerotium rolfsii*) (Chapter 11), root-knot nematodes (Chapter 12), and soil insects such as the lesser cornstalk borer (Chapter 8).

It has been demonstrated that directed band applications of granular insecticides along the row center for control of lesser cornstalk borer preserve more beneficial predacious insects and spiders than directed basal sprays or broadcast sprays of insecticides. The use of banded applications to the pegging zone and the use of granules that fall to the soil maintain beneficial insects between the rows and in the foliar canopy. When beneficial insects are preserved, fewer insecticide applications are needed. For peanut production, broadcast applications of granular nematicides or insecticides are not necessary. It should be noted that one disadvantage of granules is that they are sometimes eaten by birds. This is not a common problem, but it does occur occasionally.

Granular applications are most likely to be effective against pests that are active at or near the soil surface if the granules are not incorporated into the soil. If the pesticide is water soluble, control of pests (e.g., root worms and nematodes) that abound below the soil surface will be improved. Although a water-soluble chemical will move farther into the soil with irrigation or rain, its effectiveness is constantly reduced by dilution and chemical alteration as it moves downward. This dilution effect is one reason that pegging-time applications of water-soluble nematicides (e.g., aldicarb and ethoprop) are sometimes inadequate if they are applied without a previous preplant treatment to reduce initial nematode populations.

Chemigation

Chemigation is the application of fertilizers or pesticides through an irrigation system. The chemical is first premixed with water or crop oil plus water in some proportion in a nurse tank. The concentrate should be agitated before and during the injection of the chemical into the irrigation water line. Chemigation should not be used unless such usage is allowed by the label and unless proper antisiphoning devices are added to the injection equipment and well-head plumbing system to prevent backflow of the chemical into the water source. Protection of groundwater and surface water from chemicals should be the first order of business when chemicals are injected through irrigation systems. Groundwater contamination with pesticides has occurred most often near wells because well sites are commonly used loading sites for conventional spray operations. With chemigation, the applicator adds to the risk by injecting the chemical directly into a pipe that is linked to the water source underground or at the land surface. Recent labels have instructions about nonmixing and buffer zones from the well site (e.g., 50 or 150 feet). The actual distance is of little benefit if the loading zone is higher than the point of the well. Thus, some common sense is necessary.

> **Chemigation should not be used unless such usage is allowed by the label and unless proper antisiphoning devices are added to the injection equipment and well-head plumbing system to prevent backflow of the chemical into the water source.**

In some states (e.g., Florida), laws require that a double antisiphon device (Fig. 14.1) be placed between the injection point and the well head if toxic chemicals are to be injected. Such devices must be installed within 10° of the horizontal plane. In addition, the injection equipment should automatically cease operation when the water flow stops. A common electrical power source for the motors of the irrigation system and the injection system is one way to achieve simultaneous shutoff. When the injection equipment is powered by a shaft that operates the irrigation pump, the installation of a belt drive from the power shaft for the irrigation pump to the injection pump will also establish simultaneous shutoff. Obviously, all equipment, including hoses, should be in good and safe operating condition. The operator is responsible and liable for any adverse effects on the environment. Chemigation should not be conducted under windy conditions or near areas where people or nontarget environments will be exposed to the chemical. The end gun on a center-pivot system should be considered a point the chemical will reach. If the end gun waters a road, chemigation should not be used.

Chemigation has been used successfully with some herbicides. One advantage to applying herbicides through chemigation is that instant activation and incorporation of the chemical into the soil is achieved. Chemigation can be effective with herbicides that are absorbed by the roots of weeds. Herbicides that are absorbed by the foliage of weeds are less likely to be effective when applied through irrigation systems. Growers should consult local extension personnel about the amount of water that should accompany the herbicide. Too much water may move the herbicide below an effective zone in the soil. Soil type and tillage practices prior to chemigation will also influence soil penetration of the herbicide.

> **Chemigation should not be conducted under windy conditions or near areas where people or nontarget environments will be exposed to the chemical.**

The effectiveness of some insecticides delivered via chemigation can be improved if the insecticide is first diluted in an appropriate crop oil. Oil may reduce evaporation of the insecticide during the irrigation process and may also enhance its penetration of the insect. Because many insecticides are highly

toxic to mammals, application through an irrigation system should be conducted with strict supervision and only in isolated areas where people will not be exposed.

Chemigation with fungicides has had mixed results. For foliar disease control, it is more likely to be effective when rainfall is not excessive. Growers in Texas and Oklahoma have had satisfactory results with chemigation for the control of foliar diseases. In those areas, low rainfall minimizes fungicide runoff from the leaves. Also, leaf spot is generally less severe in the drier areas of the United States compared with the wetter southeastern peanut-growing area. Some tests in the southeastern United States have shown higher levels of leaf spot after chemigated treatments compared with levels after ground sprays with boom applicators (Tables 14.2 and 14.3). In other tests, yields increased despite the increased leaf spot, possibly because the amount of water used for chemigation was high enough to have a positive irrigation effect and therefore override the higher disease levels (Table 14.3). However, in wet years, an increase in leaf spot followed by a decrease in yield might be a more typical response. When chemigation of peanut was tested in Georgia, combining the fungicide with oil to increase the deposition of the chemical on the leaves or using low-volume, underslung booms to apply less water as a diluent increased leaf spot control. The reduced damage caused by tractors associated with chemigation compared with ground applications resulted in higher yields and reduced soilborne disease in these tests.

The wash-off effect associated with chemigation, particularly when large volumes of water are used, is helpful when it is desirable to flush a chemical through the canopy to the soil for control of soilborne diseases. In the southwestern United States, chemigation has been used successfully to control stem rot.

The decision to use chemigation for pest control should be based on many factors. There may be biological and economic benefits from chemigation in some situations. However, the potential for environmental damage is high, which in the long

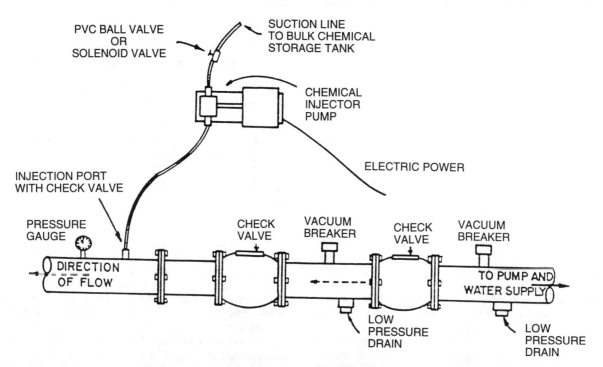

Fig. 14.1. The double antisiphon device assembly required for backflow prevention when toxic chemicals are injected into irrigation water. (Reprinted from Smajstrla et al, 1985)

Table 14.2. Effects of fungicides applied through the irrigation system (chemigation) and by ground sprayer on peanut leaf spot and yield during 3 years[a]

Treatment[b]	Yield (pounds per acre)			Infected leaves (%)			Leaves lost (%)		
	1978	1979	1980	1978	1979	1980	1978	1979	1980
None	3,194	2,392	2,805	57	62	68	20	37	40
Chemigation	4,041	4,075	4,268	40	41	38	18	24	30
Ground	3,823	3,853	3,188	21	15	61	12	09	36

[a] Adapted from Backman et al, 1981.
[b] Chlorothalonil at a rate of 2.125 pints per acre.

Table 14.3. Effect of fungicide application technique on peanut leaf spot control and peanut yield[a]

Treatment[b]	Leaf spots per 230 leaves		Yield (pounds per acre)
	Sept. 15	Sept. 25	
Chemigation	213	545	4,293
Boom sprayer	8	45	4,781
Untreated[c]	828	2,081	...

[a] Adapted from Kucharek et al, 1991.
[b] Seven applications of chlorothalonil at 2 pints per acre were made on 38.94 acres for each treatment. The chemigation treatment used 8,100 gallons per acre, and the boom sprayer treatment was at 25 gallons per acre.
[c] Three unsprayed areas 25 feet long by 20 rows wide within the boom sprayer treatment were used as untreated checks.

run could make chemigation more expensive. New laws dictate that the person who pollutes water is financially responsible for rectifying the situation. Producers should carefully consider all aspects before using chemigation.

Aerial Application of Pesticides

Pesticides are commonly applied aerially for disease and insect control (Plate 110) but are not generally used for the control of soilborne pests. Most aerial applications are made during midseason or late in the season after the vines have grown into the middles of rows. Aerial applications are most effective and drift is least likely when wind speeds are less than 5 miles per hour. Winds are often minimal during the early morning or late evening. Advantages of aerial applications include reduced workloads for the grower, reduced direct exposure to pesticides by farm crews, reduced soil compaction and vine damage, ability to spray when fields are wet, and ability to spray large areas quickly. Some of the disadvantages include reduced pesticide deposits to strategic canopy zones, more pesticide drift from the target area, and potential delays because of pilots' busy schedules. Adequate data about the effectiveness of aerial methods compared with other methods of application are not available.

Sprays are usually mixed in a nurse tank at the airstrip and pumped into the aircraft. Most spray applications for peanut are made with fixed-wing aircraft equipped with trailing-edge or under-the-wing booms (Plate 111). Variable nozzle arrangements are used, but in order to compensate for air flow from the propellers and rotor, some booms have more nozzles near the fuselage on the right wing. This arrangement tends to even out the spray pattern. With some of the new propellers, increasing the number of nozzles on the right side is not necessary.

Hollow-cone or flat-fan nozzles with specially designed diaphragms for rapid shutoff are commonly used. However, many other nozzles or devices are available. Micronair devices (Plate 112) are sometimes used, but they tend to produce smaller droplets that are prone to drift from the target area. These devices are typically used for ultra-low-volume spraying.

The pitch of nozzles and micronair rotor blades from the horizontal plane also controls droplet size because the shear force of the passing air breaks up the fluid coming through the devices. For example, nozzles pointed down or slightly backward on the boom produce droplets that are large enough to minimize drift and small enough to achieve good coverage. Nozzles pointed forward result in greatly reduced droplet size from the excessive shear forces.

Spray volumes for control of leaf spot and other foliar diseases should not be less than 5 gallons per acre. For insect control, the minimum is 3 gallons per acre. The effective spray swath of an airplane is slightly greater than the wingspan. For most planes, the swath width will range from 36 to 45 feet (11.7 to 14.6 meters), depending on the size of the plane, the boom arrangement, and the height of the flight. Spray volumes, and therefore pesticide rate, for a unit area depend upon swath width and the rate of spray from the plane. Swath widths can be marked by the grower with permanent markers. Aircraft can be equipped with devices that dictate swath width on the basis of electronic signals or mechanical devices. The use of flagmen is discouraged because they would be exposed to the pesticides.

The grower should communicate with the pilot about spray gallonage, swath width, and purpose. The purpose of the treatment in terms of pesticide rate and brand should be clearly understood by all parties. It is best to give the directions in writing. Aerial application is a complicated science and an expensive business operation. Growers should attempt to understand the basics by talking with pilots and others about variables that influence both the application techniques and legal requirements for aerial application.

Nozzles on Tractor-Mounted Spray Booms

Herbicides, fungicides, insecticides, and foliar-feed fertilizers are often applied through tractor-mounted hydraulic spray booms (Plate 113). The spray is mixed in the tank directly or pumped from a nurse tank. By pump action, usually roller, diaphragm, or piston type, the liquid spray is forced through nozzles that are strategically placed along the boom so that the spray will contact the soil or plant part that needs coverage (see section on critical spray zones).

Usually hollow-cone or flat-fan nozzles are used for spraying peanuts, but other types are available. Nozzles of the same size should be used across the boom. Every number or letter that accompanies a specific nozzle part indicates something about the characteristics of that nozzle. For example, flat-fan nozzle "8002" indicates that the spray will discharge at an 80° angle at 0.2 gallons per minute at 15 pounds per square inch (psi) if it is a low-pressure type. If a 50% increase in the sprayed rate is needed without a significant change in speed or spray pressure, an 8003 low-pressure nozzle might be used (output 0.3 gallons per minute at 15 psi). For high-pressure flat-fan nozzles, the 0.2- and 0.3-gallon-per-minute outputs are measured at 40 psi. Catalogues for nozzles will explain how the coding should be interpreted for the various nozzle parts.

For banded or broadcast herbicide sprays, flat-fan nozzles are available. Some flat-fan types (e.g., 8002E) produce an even discharge rate across the band, and some (e.g., 8002) produce a tapered rate (the rate at the outer edge of the spray pattern is less than that at the center). The even type of flat-fan nozzle is appropriate for banded herbicide applications when an even rate of chemical is required across the band. Flat-fan nozzles that produce a reduced rate on the outside of the band should be used for broadcast sprays when some overlap of

sprays from adjacent nozzles is desired to ensure complete coverage. If nontapering nozzles are used, overdoses of the herbicide might occur in the overlap area.

Fungicides used for leaf spot control can be applied with either cone or flat-fan nozzles. Although it is commonly thought that flat-fan nozzles are not effective for foliar disease control, this has been disproved for peanut leaf spot (Table 14.4). Furthermore, studies have shown that spray volumes of 15–50 gallons per acre, spray pressures of 40–300 psi, and spray speeds of 1–6 miles per hour do not influence peanut leaf spot control when the rate of the fungicide is held constant. Crop rotation and the fungicide used proved to be more important than spray pressure or spray volume (Fig. 14.2). Thus, the grower can use these variables to his advantage when designing a spray program for leaf spot. For

Table 14.4. Effects of spray nozzle type and two rates of chorothalonil on peanut leaf spot control[a,b]

Nozzle type	Nozzle tip number	Rate (pints per acre)	Leaf spots/10 leaves at 75 days	Leaflets absent/10 leaves	
				95 days	107 days
None	None	None	552	25.8	37.0
Hollow-cone	D2-25	2.125	17	0.5	5.3
Hollow-cone	D4-13	2.125	7	0.5	1.3
Flat-fan	8002	2.125	10	0.0	2.8
Flat-fan	8003	2.125	15	0.0	1.0
Flat-fan	8003	1.062	19	4.3	20.3
Flat-fan	8002	1.062	34	1.5	15.5
Hollow-cone	D4-13	1.062	28	2.5	13.5
Hollow-cone	D2-25	1.062	27	3.5	16.0

[a] Adapted from Kucharek et al, 1986.

[b] Chlorothalonil treatments were applied beginning at 39 days after planting and at 14-day intervals thereafter. Fungicides were applied in 40 gallons of water per acre at 30 pounds per square inch with three nozzles per row. Days represent the number of days after planting when assessments were made.

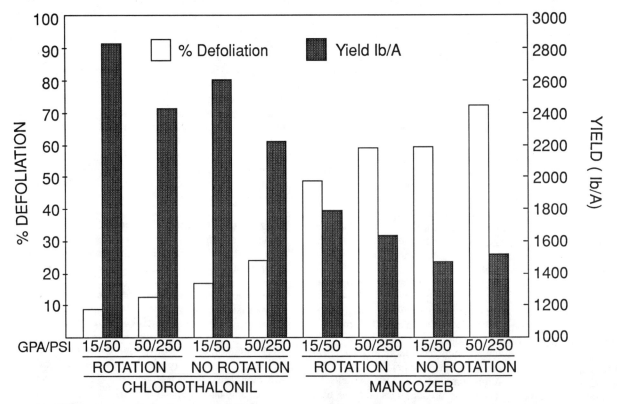

Fig. 14.2. Efficacy of chlorothalonil (Bravo 6F at 1.5 pints per acre) and mancozeb (Dithane M-45 at 1.5 pounds per acre) on control of peanut leaf spot where peanut did (no rotation) or did not (rotation) precede the crop. The fungicides were applied at different spray volumes (gallons per acre [GPA]) and pressures (pounds per square inch [psi]). Defoliation and yield were measured 119 and 134 days after planting, respectively. Disease control and yield were affected more by the fungicide used and crop rotation than by spray volume or pressure.

example, less drift will occur with lower spray pressures for which smaller pumps are adequate. The effects of these variables on other pests are not known.

Herbicides can be applied with various types of cone nozzles. The spray comes out as a truncated cone rather than as a narrow, flat fan. Cone nozzles tend to produce smaller droplets and should be used when the wind is low in order to minimize drift. Insecticides applied to the soil can be applied with flat-fan or cone nozzles. For insecticide applications to the foliage, hollow-cone nozzles are usually recommended, although the flat-fan types might be equally effective.

It is often thought that cone nozzles are more likely to deliver spray to the underside of a leaf. This may or may not be the case and depends upon leaf orientation. Nozzle arrangement is more important than nozzle type in depositing spray on the underside of the leaves, even when high spray pressures are used. Most spray is trapped on the upper surface of the leaf because that surface is directly exposed to the spray. The best way to attain coverage on the underside of a leaf is to use a systemic chemical.

The height of the boom influences the size of the area sprayed and the rate of spray that is applied. If the boom is too low and spray patterns from adjacent nozzles do not overlap, unsprayed bands along the swath will result. The amount of pesticide applied to the sprayed area will also be greater than desired because the broadcast rate is concentrated in a band. The results can be phytotoxic burns in the sprayed bands and no pest control in the adjacent unsprayed area. If for some reason the boom must be placed at a low point in relation to the canopy and 80° nozzles cannot cover the entire area, a change to 110°, flat-fan nozzles or wide-angle cone nozzles may be necessary. Conversely, if the boom is located at a point higher than the target, low-angle nozzles (e.g., 65°) are available for flat-fan or hollow-cone types.

The use of correct nozzles is a major factor in effective pest control.

A catalogue of spray nozzles and accessories (e.g., TeeJet and Delavan) should be acquired. They provide vivid pictures and useful tables that display nozzle parts and their capacities for different spray pressures and nozzle arrangements. Nozzle types and accessories are available for just about any kind of application necessary.

Nozzles are the final regulating devices for achieving desired spray rates and spray patterns. Unfortunately, many growers are not aware of the selection of nozzles and accessories because local farm suppliers seldom carry an adequate inventory of these parts. However, catalogues can be acquired and parts ordered directly from some major distributors. Growers often buy only what the local dealer is willing to stock or acquire. The use of correct nozzles is a major factor in effective pest control.

As stated earlier, sprayers can also be used for application of fertilizers. It must be remembered that fertilizers are salts and are therefore corrosive to equipment. It is always important to thoroughly clean and flush spray equipment after any use.

Air-Blast Sprayers

Air-blast sprayers use fanned air to deposit the spray across a designated swath. Relatively small sprayers can discharge a

spray across a wide swath. When this principle is combined with low spray volumes, a large acreage can be sprayed rapidly. Air-blast sprayers should not be used if a directed spray is required. They are for broadcast sprays only and are used most commonly for application of fungicides and insecticides. Adequate data about the effectiveness of these sprayers for peanut have not been gathered.

Two problems exist with the use of air-blast sprayers for crops such as peanut. First, compared with a hydraulic boom with nozzles, the spray pattern of the air blast is more likely to be distorted by wind and not reach the target evenly. For this reason, ineffective control, particularly of peanut leaf spot, has been observed with the use of these devices. This problem can be partially overcome by spraying only when the wind is calm. Second, there is a tendency to spray swaths that overextend the designed capabilities of the different machines, resulting in streaked patterns of pest control.

Air-assist sprayers that are accompanied by directed nozzles have recently been developed. The Berthoud sprayer, for example, channels high-velocity air (greater than 100 miles per hour) through a 10-inch-diameter aluminum pipe. From this pipe, the air goes into vertical polyvinyl chloride pipes with air-shear nozzles at the ends. These pipes are dropped, and the nozzles are strategically directed at the canopy to be sprayed. The spray is injected into the air-shear nozzle in advance of the discharge point. The Degania sprayer uses air fanned through a flume made from a collapsible plastic tube. The high-speed air (greater than 100 miles per hour) is then directed to holes along the bottom of the flume. Conventional nozzles are on a boom located just below the plastic tube. The fanned air forces the spray downward along the entire length of the boom. The advantages and disadvantages of these sprayers have not been completely assessed at this time, but both types have performed as well as conventional sprayers for control of white flies in peanut.

Miscellaneous Pest-Control Devices

Controlled droplet applicators (CDAs) are electrically powered, spinning disks mounted on a tool bar. Several years ago, there was considerable interest in the potential use of CDAs. It was thought that reduced spray volumes and chemical rates would be possible without loss of control of foliage pests, since the small, uniformly sized droplets would provide even coverage. However, the notion that these small, uniform droplets provide more effective suppression of plant diseases has never been supported by data. Droplet size is controlled by disks with different-sized teeth and by disk rotation speed. The centrifugal force of the spinning disk into a horizontal plane along with the small droplet size may result in streaking spray patterns, even with low wind speeds. Therefore, the effectiveness of CDAs was variable. Also, CDAs were expensive, and frequent mechanical difficulties were encountered. However, improvements have been made in the design of these devices.

Electrostatic sprayers are available but are currently used only by a few highly trained individuals for some horticultural crops. The real potential and advantages of electrostatic sprayers have not been studied in terms of pest control. The electrostatic sprayer imparts a negative charge to the spray particles, which are attracted by the positive charge of the leaves. The idea is sound, but more research is needed on peanut pest control with these devices.

Vacuum devices are being used to some extent by some growers of horticultural crops. The device sucks up insects, thereby reducing the need for insecticides. The success of these devices and their prices vary. Thus, the grower is advised to obtain a demonstration on the effectiveness of any such machine before attempting to use it for peanut pest management. Certainly, the idea is good, but again more research is needed.

Rope wick applicators have been used by a few growers of agronomic crops. With these devices, a systemic herbicide (e.g., glyphosate) is wicked along a fibrous cord to the saturation point. The wick is in a horizontal position, and as the tractor moves, the chemical is distributed directly to anything that it touches. The herbicide is thus applied to weeds that have grown above the crop canopy. Such an application is considered to be a salvage treatment where prior controls have been ineffective.

Spray Adjuvants

An adjuvant is a chemical that enhances pest control, spray operations, or environmental safety. Among the adjuvants on the market are surfactants, supplements, detergents, wetting agents, penetrants, oils, crop oils, vegetable oils, crop-oil concentrates, phytoblands, stickers, film foamers, extenders, spreaders, spreader-stickers, deposit builders, binders, thickening agents, film makers, foams, emulsifiers, dispersants, antiflocculants, stabilizing agents, synergists, sequesterents, safeners, coupling agents, cosolvents, compatibility agents, buffering agents, humectants, antifoam agents, modifiers, foam busters, and all-purpose spray adjuvants. It is often difficult to determine which type, if any, is needed for a given spray operation.

The addition of adjuvants can be beneficial for pest control, or their use can cause negative effects such as phytotoxic burns and reduced efficacy. Oil-type adjuvants that are designed to enhance penetration of weeds by the herbicide will also enhance penetration of peanut tissue by fungicides and insecticides, possibly causing burns on leaves and other tissues. A thorough understanding of the advantages of an adjuvant is necessary before it is used. Miraculous results should not be expected from addition of an adjuvant to the spray mix.

> **The addition of adjuvants can be beneficial for pest control, or their use can cause negative effects such as phytotoxic burns and reduced efficacy.**

The key to success with adjuvants is to use them infrequently, unless the pesticide label makes a recommendation or a credible source such as the extension service has clearly demonstrated success with the use of a specific adjuvant in combination with the pesticide. There are more claims made for spray adjuvants than there are adequate data for their support. Many pesticide formulations already contain certain adjuvants (e.g., spreaders, stickers, and compatibility agents).

When use of an adjuvant is being considered, the pesticide label should first be checked to see whether it suggests that a certain adjuvant would be beneficial. A label statement such as this is likely based upon data. Some labels have no information about the use of adjuvants. This could mean that the manufacturers have no data for adjuvant usage, which makes

the grower fully responsible for any incompatibility problems that might occur (e.g., settling out in the tank, nozzle plugging, phytotoxic burns, and reduction in effectiveness of the pesticides). Better control may be achieved by combining two pesticides than by adding an adjuvant. Labels should be checked for hints in this regard. Finally, ask your local extension service or the manufacturer of the pesticide for information.

Adjuvants have proved beneficial with some pesticides in certain situations. Many of the wettable powder insecticides and fungicides cover the leaves better and adhere longer if a spreader-sticker is used. This is particularly beneficial for control of foliar diseases such as peanut leaf spot and web blotch. Without it, the spray may form beaded droplets that do not adhere to the leaves. However, some labels of wettable powders recommend that an adjuvant not be used.

Many flowable pesticides used for foliar sprays do not require an adjuvant with spreading and sticking properties. In these cases, the manufacturer has incorporated adjuvants into the formulation, and the label thus recommends that additional adjuvants not be added. One reason for not adding additional adjuvants is that excessive spreading of the spray can result in faster wash off from the leaves.

For postemergence herbicides, the addition of an adjuvant with spreading properties is often recommended on the label in order that the surface of weed foliage be covered as completely as possible and the amount of penetration of the weed by the chemical be increased. Similarly, the addition of a spreading-type adjuvant may increase the uptake of an insecticide by an insect for better control. Crop-oil concentrates may increase the effectiveness of an insecticide when it is applied through an irrigation system. Conversely, a pesticide label may recommend that an adjuvant not be added to the spray mix or that an adjuvant not be added for use under certain weather conditions (e.g., rain) so that crop injury will not occur. One should remember that many herbicides and insecticides, even when used at the proper rate, may cause phytotoxic burns to the crop, and these burns may be enhanced by the adjuvant.

Some of the new pesticides that are available or in the research stage are more effective when an adjuvant is added. This is particularly true for some of the sterol-inhibiting fungicides. Because these new pesticides tend to be used at extremely low rates (ounces rather than pounds per acre), adjuvants may be helpful in reducing wash off so that the benefits of these expensive materials are maximized.

One common disadvantage of adjuvants is the enhancement of phytotoxic burns, which are most likely to be associated with the use of emulsifiable concentrate pesticides or oil-type adjuvants. These products are capable of partially damaging the cuticle (protective skin) of leaves. Adjuvants should always be used with care; no adjuvant should be used on a large scale until it has first been tried on a small area.

Calibration of Equipment and Calculation of Chemical Rate

Sprayer Calibration

Calibration is the testing and adjustment of variables such as ground speed (time), spray pressure, nozzle number, and nozzle sizes to determine their effects on the final volume discharged from the machinery. The primary objective is to relate, by ratio, the amount of spray or fumigant discharged for a known small area to the actual treated acreage. Equipment must not be leaking and must be functioning properly before calibration. Several tests may be required to correctly adjust final spray or fumigant volume.

If a major change is required in spray volume, changing the size of the nozzle tips will best meet this need. Larger or smaller nozzle tips will allow an increase or decrease in spray output, respectively, without major changes in speed or spray pressure. Manufacturers of nozzles and spraying accessories provide tables in their catalogues that relate nozzle sizes (or fumigation disks) to spray output in gallons of water per minute. For example, if doubling the output is required, a nozzle may be selected that has twice the output in gallons per minute at a given spray pressure. Increasing the number of nozzles will proportionally increase the spray output at the same spray pressure and speed. Changes in speed and spray pressure should be used for small adjustments in sprayer output.

> ## If a major change is required in spray volume, changing the size of the nozzle tips will best meet this need.

Once the spray volume (gallons per acre) is determined, the amount of chemical required (for each acre of spray volume or gallon of spray) can be calculated. Calculating the amount of chemical per gallon of water is convenient when sprayers are commonly loaded with different amounts of water to accommodate different areas. This, in turn, should reduce the amount of surplus spray. There are several basic methods of sprayer calibration.

Spraying a Known Area. This method is used commonly for calibration of aircraft and all types of ground equipment. The method is simple and takes into account all the existing nozzles or discharge devices that are in place on the equipment. It does not take into account variation that might occur between different nozzles or other discharge devices (e.g., micronair devices and CDAs). Older nozzles may be worn and may be discharging more spray than newer nozzles along the boom. When this occurs, calibration for the spray swath may be correct but the rates of chemicals applied will vary within the sprayed swath. Calibration is achieved by spraying water from the machine at the same revolutions per minute (for power-take-off pumps), pressure, and speed over an acre or a known portion of an acre that will be used for actual spraying operations. The tank is then refilled to the original mark, and the amount of water replaced is determined. If a broadcast application is made, the amount of discharged water is the spray volume for that portion of an acre (Box 14.2).

Another way to relate test-area data to an acre is to use the number of linear feet of row covered in the swath rather than the square area. If done correctly, the result will be the same. For example, a peanut crop planted on 3-foot centers will have 14,520 linear feet in one acre (43,560/3). If the boom sprays 12 rows and 1/8 of an acre is to be used for calibration, then calibration water is discharged for 151.2 linear feet. That is,

14,520 linear feet per acre × 1/8 acre = 1,815 linear feet
1,815 linear feet per 12 rows = 151.2 linear feet per row

When a spray is applied in a band, the amount of discharged water in the test is the spray volume for the geographical acre covered by the swath. However, with banded

sprays, the spray volume per treated acre in the band is greater than the area in a broadcast acre unless the spray output is reduced proportionately to the actual sprayed area. The procedure outlined in Box 14.2 can be used to calculate spray

---

Box 14.2

Determining the Spray Volume for a Portion of an Acre

Calibration of a sprayer with a boom that will cover a 12-row swath width with rows that are planted on 3-foot centers includes the following steps.

Step 1
Determine the swath width.

3 feet × 12 rows = 36 feet

Step 2
Determine the number of square feet in the portion of an acre to be used in the calibration by dividing the number of square feet per acre (43,560) by the portion of an acre to be covered in calibration (in this case 1/8 of an acre).

43,560 × 1/8 = 5,445 square feet

Step 3
Determine the distance to travel to spray 1/8 acre by dividing the area by the swath width.

5,445 square feet/36 feet = 151.2 feet

The following table gives additional distances and swaths for given portions of an acre.

Distances to be traveled for various boom widths to cover selected segments of an acre

Spray width (feet)	Linear feet to be driven to achieve:				
	1 Acre	1/2 Acre	1/4 Acre	1/8 Acre	1/16 Acre
8	5,445	2,723	1,361	681	341
12	3,630	1,815	908	454	227
14	3,111	1,556	778	389	195
16	2,723	1,361	681	340	170
20	2,178	1,089	545	272	136
30	1,452	726	363	182	91
40	1,089	545	272	136	...
50	871	436	218	109	...
60	726	363	182	91	...

Step 4
Spray the area (151 feet × 12 rows wide) at the desired tractor speed and spray pressure. Then measure the volume of water required to replace that which has been applied (for example 3 gallons). The spray volume per acre will then be the replacement water volume times eight (since 1/8 of an acre was used).

3 gallons × 8 = 24 gallons per acre

volume per treated acre as well as chemical rate. It should be remembered that boom height affects the band width and likewise the rate per sprayed area. Even if a field is sprayed with the same specifications (e.g., speed and pressure) that were used in the calibration test, the rates of both the spray volume and the pesticide will be changed in the applied area if the boom height differs from the height used in the calibration procedure.

Box 14.3

The Catch Method

Example: A spray boom has 12 nozzles, 1 foot apart, along a 12-foot boom. The desired ground speed is 4 miles per hour. If the sprayer is stationary and a power-take-off pump is used to supply the pressure, the revolutions per minute should be set to match operation at 4 miles per hour.

Step 1

Selected nozzles along the boom (in this case three nozzles) may be used to collect the spray of water for 1 minute at the prescribed settings. An average of 37 fluid ounces per nozzle per minute is measured. Multiply the average nozzle output by the number of nozzles for total output.

$$37 \times 12 = 444 \text{ ounces per minute}$$

Step 2

Divide the total number of ounces per minute by 128 (the number of fluid ounces in 1 gallon) to get the number of gallons per minute.

$$444/128 = 3.5 \text{ gallons per minute}$$

Step 3

Calculate the number of square feet covered by the swath in 1 minute at 4 miles per hour (88 feet per minute are covered at 1 mile per hour).

$$4 \times 88 = 352 \text{ linear feet of forward travel}$$

and

$$352 \text{ feet} \times 12\text{-foot swath width} = 4,224 \text{ square feet}$$

Step 4

Now the actual spray volume per acre can be calculated, because it is known that the sprayer applies 3.5 gallons per 4,224 square feet and there are 43,560 square feet in an acre.

$$3.5 \text{ gallons}/4,224 \text{ square feet} = X \text{ gallons}/43,560 \text{ square feet}$$

$$X = (3.5 \text{ gallons} \times 43,560 \text{ square feet})/4,224 \text{ square feet}$$
$$= 36 \text{ gallons}$$

Therefore, 36 gallons per acre will be applied by the sprayer at 4 miles per hour.

Catch Method. The catch method is most useful for testing boom applicators with nozzles. The test consists of collecting water sprayed from a select number of nozzles on the boom for a known period of time at the prescribed setting of spray pressure, revolutions per minute, and ground speed (1 mile per hour = 88 feet per minute). The procedure for calculating the actual volume of spray per acre with the catch method is outlined in Box 14.3.

If minor changes in spray volume are needed, spray pressure or spray speed may be changed accordingly. If a major change in spray volume is needed, larger or smaller nozzles are used or the number of nozzles is increased or decreased as required. Obviously, the catch method is better than the method of spraying a known area for determining the output of individual nozzles. If all nozzles are checked, those that are obviously worn, as indicated by higher outputs, or otherwise defective should be replaced.

Calculation of Chemical Formulation. The calculation of the amount of chemical to be added to the spray tank is made after the amount of spray volume of water is determined (Boxes 14.2 and 14.3). If 125 gallons of spray is to be mixed and the spray volume was determined by calibration to be 25 gallons per acre, the amount of chemical to be added to the tank should be the rate per acre times five for the 5 acres to be sprayed. If the rate per acre is 2 pounds of formulated product, then

$$2 \text{ pounds} \times 5 \text{ acres} = 10 \text{ pounds added to the spray tank.}$$

Calibration of Granular Applicators

Mathematical relationships similar to those employed for calibration of sprayers are used to calibrate granular applicators, except that the weight of the chemical, not the volume, is used. With granules, calibration will determine the amounts of the formulation and chemical ingredient at the same time. In tests for broadcast applications, the granules that are discharged for a known portion of an acre are collected. Those granules must be weighed and related, by ratio, to an acre. For example, if 100 pounds of granules is required for an acre (43,560 square feet) and 1.38 pounds was caught for 600 square feet of test, the applicator settings and speed are correct.

$$(1.38 \text{ pounds}/600 \text{ square feet}) \times 43,560 \text{ square feet}$$
$$= 100 \text{ pounds per acre}$$

If more or less than the desired amount of granules is collected, the ground speed or the applicator setting may be altered, unless the applicator operates independently of these two variables.

For band applications of granules, the same mathematical relationship can be used on a square footage basis, taking into account that the material will actually be applied along a band. However, the same results can be achieved by establishing a ratio of row footage in the test area to the row footage in an acre. For example, the row-footage calibration for 100 pounds of granules to be used along 14,520 row feet per acre (based on 3-foot row centers) is as follows:

$$X \text{ pounds}/100 \text{ feet of row} = 100 \text{ pounds}/14,520 \text{ feet of row}$$

$$X \text{ pounds} = (100 \text{ pounds} \times 100 \text{ feet of row})/14,520 \text{ feet of row}$$
$$= 0.69 \text{ pounds per 100 feet of row should be collected}$$

The applicator setting may be raised or lowered if the amount weighed is too low or high, respectively. Also, decreased or increased ground speed will raise or lower the rate, respectively. Rotor speed can be adjusted on some electrically driven units to enhance precision at low and high rates. Adjustments should be made and testing should continue until the correct output is achieved.

Calibration of Liquid Fumigant Applicators

The same mathematical relationship used for sprayers may be used for liquid fumigant applicators. With liquid fumigants, it is critical that protective clothing and an appropriate face mask be worn during the calibration process because of the toxicity and vaporization of the liquid. The person collecting the fumigant must be upwind from the flow tubes. Appropriate gloves without holes are also essential for this activity. The flow rates (fluid ounces per linear foot), usually for 1,000 feet of row, for these products are printed on the label. The forward speed must first be determined and from this, the number of feet to be traveled in 1 minute. Then the liquid can be caught for 1 minute while the tractor is stationary and operating at the correct revolutions per minute. This will give the flow rate per tube per linear foot of row. Whether a gravity-feed system or a pump system is used for the fumigant, the rate of application is inversely proportional to speed.

> With liquid fumigants, it is critical that protective clothing and an appropriate face mask be worn during the calibration process because of the toxicity and vaporization of the liquid.

If the amount of fumigant is significantly high or low compared with the desired amount for a known distance at a known speed, smaller or larger disks, respectively, can be used in the distribution manifold to achieve the desired amount. Where minor changes are needed to obtain the correct rate, tractor speed can be increased or decreased.

Calibration for Chemigation

This calibration requires that the number of acres to receive irrigation (including end-gun coverage) and the speed of the irrigation device be known. The rate of the premixed chemical in the nurse tanks that is to be injected into the water output line must be constant from the beginning to the end of the operation. Likewise, irrigation nozzles on the system must deliver the same amount of water to all areas of the field. With center-pivot systems, this requires proportionately larger nozzles from the center to the outer end of the system because the system is moving faster at the outer end. The amount of discharged water can be measured by placing rain gauges or pans along a transect from one end of the system to the other. A sample calculation of calibration for a center-pivot irrigation system is given in Box 14.4.

Selected References

Backman, P., Crawford, M. A., and Rochester, E. W. 1981. Application of fungicides to peanuts through the irrigation system. Highlights of Agricultural Research, vol. 28, no. 2. Auburn University, Auburn, AL.

Brenneman, T. B., and Sumner, D. R. 1989. Effects of chemigated and conventionally sprayed tebuconazole and tractor traffic on peanut diseases and pod yields. Plant Dis. 73:843-846.

Dickson, D. W., and McSorley, R. 1987. Standardization of nematicide application rates. Ann. Appl. Nematol. 1:1-5.

Kucharek, T. A., Cullen, R. E., Stall, R. E., and Llewellyn, B. 1986. Chemical control of foliar diseases of peanuts, peppers, and onions as affected by spray nozzle types, nozzle orientations, spray inter-

Box 14.4

Calibrating for Chemigation with a Center-Pivot Irrigation System

This example is for a center-pivot irrigation system with a radius of 1,325 feet (distance from the center to the end of the area reached by the last nozzle or end gun, if it is used).

Step 1

Determine the circumference (total distance around the outer edge) of the circle made by the system when it has traversed 360° with the equation $2\pi \times$ radius or

$$2 \times 3.14 \times 1,325 \text{ feet} = 8,321 \text{ feet.}$$

Step 2

Determine the actual speed of the irrigation system at the setting for the desired speed. The distance traveled by the outermost wheels for 10 minutes can be measured two or three times to get an average. If the unit moves 73.5 feet in 10 minutes, the time required to traverse 8,321 feet is

$$73.5 \text{ feet/10 minutes} = 8,321 \text{ feet/X minutes}$$
$$X = 10 \text{ minutes} \times 8,321 \text{ feet/73.5 feet}$$
$$= 1,132 \text{ minutes.}$$

Then 1,132 minutes/60 = 18.9 hours to traverse a complete circle, 18.9 hours/4 = 4.72 hours to traverse one quadrant, or 1,132 minutes/4 = 283 minutes to traverse one quadrant.

Step 3

Determine the output per minute of the premixed chemical from the injection pump at a known setting. If the pump capacity in gallons per hour is known and the output of the pump is linear with the settings, the setting can be changed between trials until the correct amount is achieved. The flow rate with the injection pump should be tested with the actual spray mix in the nurse tank since concentrated mixes are more likely to alter flow rates from pumps than more dilute mixes. If 43.5 gallons of spray mix is needed to cover one-quarter of a revolution in the 283 minutes mentioned above, the amount of concentrated spray mix to be injected by the pump in 1 minute must be determined by

$$43.5 \text{ gallons per 283 minutes} = X \text{ gallons per minute}$$
$$X = (43.5 \text{ gallons per minute})/283 \text{ minutes}$$
$$= 0.154 \text{ gallons per minute} \times 128 \text{ ounces per gallon}$$
$$= 19.66 \text{ ounces per minute.}$$

vals, and adjuvants. Plant Dis. 70:583-586.

Kucharek, T., Shokes, F. M., and Gorbet, D. 1991. Considerations for spraying foliar fungicides to control plant diseases as exemplified by studies in Florida from 1968 to 1989 on the control of peanut leaf spot. Univ. Fla. Bull. 269.

Phipps, P. M. 1990. Control of Cylindrocladium black rot of peanut with soil fumigants having methyl isothiocyanate as the active ingredient. Plant Dis. 74:438-441.

Smajstrla, A. G., Harrison, D. S., Becker, W. J., Zazueta, F. S., and Haman, D. Z. 1985. Backflow prevention requirements for Florida irrigation systems. Univ. Fla. Circ. 217.

Smith, J. W., and Jackson, P. W. 1975. Effects of insecticidal placement on non-target arthropods in the peanut ecosystem. Peanut Sci. 2:87-90.

T. D. Hewitt
North Florida Research and Education Center
University of Florida, Marianna

F. M. Shokes
North Florida Research and Education Center
University of Florida, Quincy

CHAPTER FIFTEEN

Economics and Impacts of Decision Making in Peanut Health Management

As the preceding chapters illustrate, managing the health of a crop involves many factors. Many decisions must be made before a particular cropping system is implemented. We have seen, for example, the need for crop rotation in the management of many diseases. Stating that crop rotation is essential is one thing; determining the crops for rotation that will allow a sustainable yield and profit is quite another. Economic decision making is a vital part of the management of any crop, and peanut is no exception. Regardless of the ability of the producer to grow and maintain a healthy crop, a profit must be made to stay in business. Profitability is an important factor that must be considered when any management decision is made. Production decisions and profitability must not be treated independently.

To cope with the risks and uncertainties associated with agricultural production, good records are needed to provide proper information for planning and monitoring operations. Daily management analysis has become an important part of the farming operation. Government requirements must be closely followed by peanut producers because of the quota production system of pricing (Box 15.1). Management decisions for a crop of quota peanuts with a fixed value per ton may differ from the decisions made for a crop of additional peanuts, the value of which is dependent on supply and demand.

> **The question, Does it pay? must be answered.**

Because of narrow profit margins, making small changes in the current operation or employing new production plans may be critical to the economic well-being of the peanut producer. The economic aspects of changes in farming practices must be a part of the decision-making process. The question, Does it pay? must be answered. Various costs (cash, social, and envi-

ronmental) and benefits must be analyzed and should be listed for consideration. Once costs are determined, benefits need to be weighed against them. Monetary returns must be calculated along with other nonmonetary benefits. Once costs and benefits are analyzed, decisions may then be made. If the projected

---

Box 15.1

Terms Associated with Peanut Management

additionals peanuts grown without poundage restrictions; supported at a lower price than quota peanuts

cost-benefit analysis an evaluation of the associated costs and benefits derived from a business venture

enterprise budget a detailed list of all estimated expenses and revenues associated with a specific enterprise

fixed costs costs that do not vary and cannot be avoided, regardless of the amount of a crop produced (e.g., land and machinery)

input a component of production (e.g., land, labor, seed, or chemicals)

operating expenses expenses incurred for crop production within a production cycle (e.g., seed and fertilizer)

partial budget a budget that includes only the revenue and expense items that would change as a result of some type of change in the business production technique

quota peanuts with the price fixed by the poundage level allowed under a government program

variable costs costs that change with the amount of crop produced and that are avoidable if production ceases (e.g., electricity to run an irrigation pump)

cost of a practice does not appear profitable on paper, then the producer needs to consider the alternatives.

The economic aspects of peanut production need to be discussed along with the role and importance of farm management and decision making. The concepts of enterprise budgeting and partial budgeting must be considered. These two

The concepts of enterprise budgeting and partial budgeting must be considered.

management tools can be very helpful in assisting peanut producers in their production planning and day-to-day decision making. The management steps outlined here are similar for any crop, but they will be discussed in relation to the economics of peanut health management. Some of the terms that are necessary for a discussion of peanut management are defined in Box 15.1.

Importance of Farm Management

Farm management is a very important component of the overall farming operation because management decisions determine how effectively and profitably the land, labor, and capital resources are used. Often the differences in profit levels can be attributed to management. Challenges to successful farm management include dwindling energy resources, government policy changes, tax laws, food-quality concerns, and restrictions in the use of certain agrochemical inputs. Successful farm managers plan, evaluate, and choose among the available options for proper use of available resources to accomplish the farm goals.

Planning is a very important part of farm management. It forces producers to evaluate their limited resources and to realize that all goals will not be accomplished. Farm planning is an ongoing process as new problems and opportunities arise and new information and technology become available. The list in Box 15.2 gives some idea of the scope of farm management, which is concerned with all the technical areas of agriculture that afford profitability.

Management includes decision making, action taking, and evaluation. Successful peanut producers run their farms economically, using available resources to achieve the desired goals. Decisions must be made that will pay off over the long term. To engage in successful peanut production, the grower must perform most or all the decision-making and planning functions listed. To do this, some knowledge of the various steps in decision making is necessary. The ability to make and improve on decisions over time is crucial to farm survival.

Budgeting

Farm managers have different tools to assist in decision making. Proper use of these tools will help peanut producers improve the accuracy of their decisions. Both break-even analysis and cost-benefit analysis should be performed. Break-even analysis is often used to consider how price and/or yield changes might affect the selection of alternative enterprises, production methods, or inputs. In cost-benefit analysis, the costs must be determined and weighed against the potential benefits. Both economic and environmental costs and benefits must be considered on today's farm.

The two types of management tools that are most helpful to peanut producers are enterprise and partial budgeting. Farm budgeting procedures can be used to compare farming alternatives on paper before investing in an idea. Enterprise budgeting may be used to evaluate expenses, potential revenue, and profit or loss for a particular commodity over a given time. Managers may also perform a whole-farm analysis to look at the potential for the total operation. In some cases, many parts of the farm business may not have to be considered when the consequences of a particular decision are analyzed. For this type of analysis, partial budgeting is appropriate.

Enterprise Budgeting

Projections of annual costs and returns for an enterprise are called enterprise budgets. An enterprise budget contains all the income and expenses associated with a single enterprise, including operating and fixed costs. Operating costs are those that are directly associated with a specific enterprise (for example, the cost of the peanut seed) and are relatively easy to estimate. Fixed costs are those that are associated with more than one enterprise and do not change with the level of production (for example, the cost of insurance on farm equipment). Fixed costs must be allocated to all associated enterprises and are not as readily estimated as operating expenses.

Developing an Enterprise Budget. The first step in developing an enterprise budget is to identify the enterprises on the farm. An enterprise is generally defined as a crop or type of livestock produced for profit. Growers may even identify individual fields of the same crop as separate enterprises. Growers who develop separate budgets for each field may sometimes find that the differences are not significant enough to warrant the required paperwork. However, other farmers who develop one budget for each commodity find that the information generated is not detailed enough to make certain production decisions.

A farmer might divide a peanut crop into two enterprises, quota and additionals (Box 15.1), rather than treating the entire crop as one enterprise. The price for additionals is not fixed and will often be lower than the quota price. A peanut grower must decide whether to contract additionals at a fixed price or take a chance on getting a higher price. The expected return for the crop will weigh heavily in decisions related to the use of various inputs. By subdividing the same commodity into several enterprises, the farmer can monitor crop per-

Box 15.2

Farm-Management Considerations

- What crop or crops should be produced?
- How much land should be used for a specific crop?
- What method of production should be used?
- Which agrochemicals are necessary?
- When should the necessary inputs be applied?
- Where should the inputs be purchased?
- When and where should the product be sold?

formance on a field-by-field basis and pinpoint problems as they occur.

In general, enterprise budgets are constructed from whole-farm records by allocating the income and expense items for the whole farm to individual enterprises. If records are not available or the farm manager is interested in a new commodity for the farm, information is usually available from the state land grant university's cooperative extension service. Extension economists often develop enterprise budgets as guides for farm planning, and production specialists make recommendations in these production guides for various commodities.

A budget is a systematic listing of income (revenue) and expenses for a production period. Two general types of costs, variable and fixed, make up the total cost of producing any farm commodity. Variable costs change with the decisions made about how many units of the various inputs to include in the operation. Variable costs are the costs of inputs that are used up during one year's operation or during one production period. Fixed costs are associated with buildings, machinery, and equipment and should be prorated over a period of years. Fixed costs are not affected by short-term enterprise decisions.

The enterprise budget will indicate the profitability of

Table 15.1. Estimated revenue and costs of producing 1 acre of quota peanuts in northern Florida, 1994

Item	Unit	Quantity	Price	Value
Revenue				
Quota peanut receipts	ton	1.75	$700.00	$1,225.00
Variable costs				
Seed	pound	90.00	0.85	76.50
Fertilizer				
Nitrogen (N)	pound	15.00	0.28	4.20
Phosphate (P_2O_5)	pound	45.00	0.25	11.25
Potash (K_2O)	pound	90.00	0.14	13.50
Lime (spread	ton	0.50	24.00	12.00
Gypsum (spread)	hundredweight	6.00	2.25	13.50
Herbicide	acre	1.00	54.00	54.00
Insecticide	acre	1.00	40.00	40.00
Fungicide	acre	1.00	45.00	45.00
Nematicide	acre	1.00	90.00	90.00
Spraying (air)	application	4.00	5.00	20.00
Scouting fee	acre	1.00	4.50	4.50
Tractor (135 hp)	hour	3.00	8.90	26.70
Tractor (55 hp)	hour	2.25	3.50	7.88
Equipment	hour	5.25	3.35	17.59
Truck, pickup	mile	40	0.14	5.60
Drying and cleaning	ton	1.75	40.00	70.00
Peanut commission	ton	1.75	2.00	3.50
Hired labor	hour	4.00	5.00	20.00
Land rent	acre	1.00	25.00	25.00
Interest on variable costs[a]	$	560.72	0.05	28.04
Total variable costs				588.76
Fixed costs				
Tractor (135 hp)	hour	3.00	12.00	36.00
Tractor (55 hp)	hour	2.25	6.45	14.51
Truck, pickup	mile	40	0.17	6.80
Equipment	hour	5.25	10.10	53.03
Total fixed costs				110.34
Total costs				699.10
Returns above cash and fixed costs (returns to labor and management)				525.90

[a] 10% for 6 months.

Break-even peanut prices at various yields

Yield (lb/acre)	Price for variable costs ($/lb)	Price for total costs ($/lb)
2,000	0.29	0.35
2,500	0.24	0.28
3,000	0.20	0.23
3,500	0.17	0.20
4,000	0.15	0.17
4,500	0.13	0.16
5,000	0.12	0.14

Your break-even price = your cost/your yield.

producing an acre of peanuts. The revenue and expenses estimated in the enterprise budget are those expected for use of particular production practices. The budget covers the entire production cycle, indicating resources required, end products, costs incurred, and prices received.

Two sample peanut budgets are given. One is an estimate of expenses and revenue for quota peanuts in northern Florida, and the other is an estimate for production of additional peanuts. These budgets are composite projections made by using extension recommendations and actual production records of selected producers. A table is included for each budget to show the prices that would be needed to break even at various yields. A grower should never produce a crop if the budget does not project a positive return over the variable costs. In these sample budgets, note that agrochemical costs are grouped into specific types: herbicides, insecticides, and fungicides. Extension budgets do not usually list specific chemical inputs because of problems with making specific

recommendations of one compound over another. These costs are derived from a mix of agrochemical inputs commonly used by farmers and recommended by extension specialists.

If a producer is contemplating certain changes, the aftereffects of these changes can be evaluated by using the enterprise budget. For example, a peanut producer may estimate the effect of land rent or quota rent on the returns. The effects of yield or price changes can also be quickly analyzed by using part of the enterprise budget.

Enterprise budgets can summarize costs and returns, inputs and production, and timing of resource use for a particular farm activity. Since enterprise budgets contain both ownership and operating costs, the budgets can be used in long-term planning to 1) provide data for whole-farm planning, 2) estimate potential income for particular farm situations, 3) estimate the size of a farm needed to earn a specified return, 4) estimate cash flows during the year, and 5) estimate the cost of production for different agricultural products.

Table 15.2. Estimated revenue and costs of producing 1 acre of additional peanuts in northern Florida, 1994

Item	Unit	Quantity	Price	Value
Revenue				
Additional peanut receipts	ton	1.25	$350.00	$437.50
Variable costs				
Seed	pound	80.00	0.85	68.00
Fertilizer				
Phosphate (P_2O_5)	pound	30.00	0.25	7.50
Potash (K_2O)	pound	90.00	0.15	13.50
Lime (spread)	ton	0.50	24.00	12.00
Gypsum (spread)	hundredweight	7.00	2.25	15.75
Herbicide	acre	1.00	38.00	38.00
Insecticide	acre	1.00	30.00	30.00
Fungicide	acre	1.00	32.00	32.00
Tractor (135 hp)	hour	2.50	8.90	22.25
Tractor (55 hp)	hour	2.00	3.50	7.00
Equipment	hour	4.50	3.35	15.08
Truck, pickup	mile	40	0.14	5.60
Drying and cleaning	ton	1.25	40.00	50.00
Peanut commission	ton	1.25	2.00	2.50
Hired labor	hour	2.50	5.00	12.50
Land rent	acre	1.00	15.00	15.00
Interest on variable costs[a]	$	346.68	0.05	17.33
Total variable costs				364.01
Fixed costs				
Tractor (135 hp)	hour	2.50	12.00	30.00
Tractor (55 hp)	hour	2.00	6.45	12.90
Truck, pickup	mile	40	0.17	6.80
Equipment	hour	4.50	10.10	45.45
Total fixed costs				95.15
Total costs				459.16
Returns above cash and fixed costs (returns to labor and management)				−21.66

[a] 10% for 6 months.

Break-even peanut prices at various yields

Yield (lb/acre)	Price for variable costs ($/lb)	Price for total costs ($/lb)
2,000	0.18	0.23
2,500	0.14	0.18
3,000	0.12	0.15
3,500	0.10	0.13

Your break-even price = your cost/your yield.

Sample Budgets for Peanuts. The sample budgets in Tables 15.1 and 15.2 are estimated for yields of 3,500 pounds per acre for quota peanuts and 2,500 pounds per acre for additional peanuts. Inputs are projected to be less for production of additional peanuts since potential revenue is usually less. Use of more intensive production techniques, including irrigation, is assumed for the higher-value quota peanut crop. A price of $0.20 per pound would be needed just to break even with a yield of 3,500 pounds per acre for quota peanuts. Only $0.14 per pound would be necessary to break even with a 2,500-pound-per-acre yield of additional peanuts, and this is less than the anticipated price in the sample budget of Table 15.2.

Partial Budgeting

Producers constantly consider adjustments in their farming practices as input prices, output prices, and technology change. Ways to increase profit through alternative methods and production strategies are continually sought. Partial budgeting is a simple analytical technique for determining the profitability of small and moderate changes to the farm production process.

Partial budgeting permits organized evaluation of how a change in the farm operation will affect profit. Only factors that are directly affected by the proposed change are included in the partial budgeting analysis. Partial budgeting determines immediately whether an alternative might add to the overall profitability of the farm and can be used to weed out changes that are clearly unprofitable before more complex whole-farm analysis is undertaken.

Partial budgeting is especially useful when particular types of management changes are analyzed. Such changes might include all of the factors listed in Box 15.3.

The format for creating a partial budget is presented in Table 15.3. The benefits from a change in the farm business are weighed against the costs incurred by the change. The costs in column A consist of the added costs that will be incurred by the change and the reduced income that will result from the change. The benefits in column B are the opposite of the costs. Benefits include reduced costs of production and increased income. If total benefits are greater than total costs, then the proposed change will be more profitable than the existing situation. The opposite is true if the costs are larger than the benefits.

Accurate cost and income information is needed for partial budgeting. Calculating accurate cost and income information requires accurate price and quantity estimates. Prices and quantities are usually not known with certainty. Estimates can be made more accurately, however, with careful attention to detail and with good research on the problem at hand. Yield estimates can be obtained from historic farm records, university extension publications, and research publications. Prices are more difficult to estimate, although market news services, U.S. Department of Agriculture outlook estimates, and the futures market all give indications of coming price trends.

Accurate cost and income information is needed for partial budgeting.

The format presented in Table 15.3 illustrates that increased profits are possible when a change in the farm business results in some type of added income and/or reduced cost. The partial budgeting analysis will assist in assessing the change in profitability resulting from the use of an alternative.

Two examples are included to illustrate uses of partial budgeting. The examples, a change in fungicide used and the choice a leaf spot spray program, are typical of farm decisions concerning peanut production. The important concept in partial budgeting is that only those revenue and expense items that will actually change as a result of the alternative action should be included.

Example 1: Changing to an Alternative Fungicide. A peanut farmer is considering alternative methods to reduce production costs. Information on the results of research concerning various peanut diseases and pesticides has been given at a production meeting. The research results in Table 15.4 indicate that use of fungicide CO could increase yields by 400 pounds per acre compared with fungicide AB, the one now used by the farmer.

The producer currently owns the equipment needed and figures that all other production costs would be unchanged. Costs will increase, but income should also increase. Estimating the price for quota peanuts at $0.35 per pound, the difference in production costs would be $19 per acre ($57 − $38). The added income is $112 per acre.

Box 15.3

Changes in Management Practices for which Partial Budgeting Is Useful

- Substituting a new enterprise for part or all of an existing enterprise
- Substituting one input for another
- Increasing or decreasing the level of input use
- Increasing or decreasing the size or scale of an enterprise
- Purchasing a machine or some other asset specific to an enterprise

Table 15.3. Partial budgeting format

Column A	Column B
Added costs	Reduced costs
Reduced incomes	Added income
Total (Column A)	Total (Column B)
Change in profitability = B − A	

Table 15.4. Fungicide test results

Fungicide	Yield (lb/acre)	Fungicide cost ($/acre)
None	2,100	0
AB	2,800	38
CO	3,200	57
EE	2,750	35
FD	2,810	40
DD	2,675	45

The increase in income is greater than the increase in costs by $93 per acre (Table 15.5); thus, changing to the alternative fungicide would be advisable. A change in the price of peanuts or the cost of the fungicide would change the outcome. Producers must be careful in selecting the costs and prices that are used in a partial budget analysis.

Example 2: Choosing Between a Calendar Spray Schedule and a Spray Advisory for Leaf Spot Fungicide Applications. A Virginia farmer is planning his peanut production. Research data has been presented that compares a calendar-based leaf spot fungicide schedule with spray applications scheduled by using the Virginia Leaf Spot Advisory. When the advisory system was used, an average of 4.3 fewer sprays per season were applied over 4 years. The producer knows that each application costs $9.50 per acre in tractor, labor, material, and equipment. The costs of the materials must be obtained and figured into the costs of the spray programs. The information shown in Table 15.6 gives the various yields and costs, and Table 15.7 gives the effect on profitability expected from the change in practice. The producer figures the price of peanuts will be $0.35 per pound and needs to analyze the value of the different fungicide spray programs.

In this particular example, the most economical program would be the advisory program, which gives a benefit of $36.30 per acre over the 14-day program. If a producer has his fields scouted for disease problems, the cost of the scouting program should also be factored into the cost side of the partial budget.

Peanut growers could also adjust the partial budget to account for the risk involved in production. Yields, prices for inputs, and market prices can be accounted for by changing the partial budget. The budget can then be recalculated to determine whether the proposed changes remain profitable under conditions that vary from the expected.

Partial budgeting is a simple, accurate tool for the analysis of small- to medium-sized changes in the farm business. Only the factors affected by the proposed change are included in the analysis. Furthermore, once the net profitability of a proposed change is known, calculating the break-even prices above or below which the proposed change may be profitable is relatively easy. Partial budgeting is a powerful tool that every peanut farmer should employ in the analysis of alternatives to current farming practices.

Other factors must be considered when such decisions are made. In some production areas, use of a particular fungicide for a specific disease might increase the potential for another disease. For example, in Virginia, use of chlorothalonil for leaf spot control might increase the chances for Sclerotinia blight, a soilborne disease (Chapter 11).

Leaf spot forecasting programs similar to that used in Virginia are now in use in the southeastern peanut-growing regions. Growers can develop partial budgets to determine the profitability of using such programs. These models make possible decisions that are not only profitable, but also environmentally sound. Use of some advisory systems requires the purchase of equipment such as computer-controlled weather stations. The costs of such systems must be included as added costs and could change the profitability during the year of purchase. However, the benefits must be considered over the long term since the system could possibly be paid for by pesticide application cost savings during the first year.

Other decisions may also be made, on the basis of accurate farm records, that will increase the profitability of peanut production. These include the elimination of unnecessary inputs or a decrease in the use of certain inputs. Some examples of these are listed in Box 15.4. While certain changes in practices may cause some inconvenience, they might actually result in cost savings and added benefits to the environment.

Table 15.5. Partial budget for the decision to change to an alternative fungicide

Column A		Column B	
Added costs	$19	Reduced costs	0
Reduced income	0	Added income	$112
Total	$19	Total	$112
	Change in profitability (B − A) = $93		

Table 15.6. Comparison of fungicide spray programs for control of leaf spot on Florigiant peanut[a]

Spray program	Average number of applications	Pod yield (lb/acre)	Treatment cost
14-day schedule	6.8	3,525	$64.60
Advisory	2.5	3,538	$23.75
Untreated check	0	2,773	0

[a] Adapted from Phipps and Powell, 1984.

Table 15.7. The fungicide program decision

Column A		Column B	
Added costs	0	Reduced costs[a]	$40.85
Reduced income[b]	$4.55	Added income	0
Total	$4.55	Total	$40.85
	Change in profitability (B − A) = $36.30		

[a] $64.60 − $23.75.
[b] 13 pounds at $0.35 per pound.

Box 15.4

Changes in Farming Practices for which Partial Budgeting Is Useful

- Use of certain herbicides only on specific fields or parts of fields where they are needed, rather than on every part of every field. Such need can be determined by making accurate weed maps.

- Use of soilborne fungicides only on fields or portions of fields where such use is required. Accurate records must be kept for each field to determine the specific problems that occur and their locations. For example, if a disease occurs only in a low, wet area, that may be the only area where treatment is necessary, or it might be feasible to avoid planting in that area.

- Not planting fields or portions of fields on which production is likely to be marginal, for example, wet areas in which plant growth may be marginal and weed growth excessive. Not planting too close to the edge of some fields will avoid wasting inputs on plants that will be competing with trees for light and moisture.

These are just some of the types of decisions that will have to be made in the future by peanut growers. Such factors may not have been given much consideration by growers in the past because of the inconvenience of implementation. However, growers today have a greater awareness and concern for the environment. Also, there are major efforts being made toward a more sustainable agriculture. These efforts are resulting in closer scrutiny of on-farm inputs into production. Cost-benefit analysis has become increasingly important at the farm level. The use of partial budgets allows the determination of costs and monetary benefits.

Partial budgeting does not take into account the additional benefits that might be derived from changes in farm practices. For example, the decision to use a spray advisory system, such as that illustrated in Tables 15.6 and 15.7, will benefit the environment by decreasing the amount of pesticide used. Another benefit might be fewer trips over the field if a ground-type spray applicator is used. Less field traffic results in less plant injury and could reduce the amount and severity of some soilborne diseases (Chapter 11). Therefore, the decision to use the advisory system results in lower inputs without decreased yields while maintaining the sustainability of the production system in other ways as well. It is hoped that involving budgeting in the decision-making process will make producers more knowledgeable as they strive toward the wise use of available resources in their attempts at peanut health management.

In summary, applying economic principles to decision making is a very real and important part of peanut health management. Producers must be more than just production conscious. Any operations or changes undertaken on the farm must show a profit. If a plan or idea does not show a profit or cash flow on paper, producers should look at another alter-native. The bottom line to successful peanut production or any other farming venture is net income. The greater the net income, the more likely the farm will continue in existence.

Producers need to understand the importance of peanut health management in order to make prudent decisions. Many farm-management tools are available for making production decisions. Enterprise and whole-farm budgets should be developed to help in deciding what and how much to plant. Partial budgeting can be used to assist farmers when they are looking at alternative production methods and making vital production decisions. Risk and uncertainty are prevalent in agriculture today, and to survive producers need to understand the concepts presented. Wise planning and decision making are necessary for peanut producers to accomplish the goal of good peanut health management for greater sustainability on the farm.

Selected References

Calkins, P. H., and Pietre, D. D. D. 1983. Farm Business Management: Successful Decisions in a Changing Environment. Macmillan, New York.

Castle, E. N., Becker, M. H., and Nabor, A. G. 1987. Farm Business Management: The Decision Making Process. Macmillan, New York.

Ford, S. 1989. Analyzing changes in the farm business: Partial budgeting. Univ. Fla. Ext. Circ. 837.

Kadlec, J. E. 1985. Farm Management: Decision, Operation, Control. Prentice Hall, Englewood Cliffs, NJ.

Phipps, P. M., and Powell, N. L. 1984. Evaluation of criteria for the utilization of peanut leafspot advisories in Virginia. Phytopathology 74:1189-1193.

Westberry, G. O. 1979. Partial budgeting. Food Resour. Econ. Dep. Univ. Fla. Staff Pap. 130.

Index